Case Histories in Construction Law

A Guide for Architects, Engineers, Contractors, Builders

William Jabine, J.D.
Member of the New York Bar
Contributing Editor, *Actual Specifying Engineer*

Sponsoring Editor, Don DeMicheal
Editor, *Actual Specifying Engineer*

Cahners Books

Division of Cahners Publishing Company, Inc.
89 Franklin Street, Boston, Massachusetts 02110

International Standard Book Number: 0-8436-0113-2
Library of Congress Catalog Card Number: 72-81985
New material copyright © 1973 by Cahners Publishing Company, Inc.
© 1971, 1970, 1969, 1968, 1967, 1966, 1965, 1964, 1959 by Cahners
Publishing Co., Inc.
Printed in the United States of America.
The Maple Press Company, York, Pennsylvania, U.S.A.

Contents

Introduction

The case histories in this volume first appeared in the "Engineers at the Bar" department of *Actual Specifying Engineer Magazine.* They record the decisions of courts in all sections of the United States and cover a wide variety of disputed situations in the country's construction industry. All but one of the cases were published in the magazine during the last seven years. The exception is the first case, which was published in the January, 1959, issue. It is included here because it is a landmark case still cited as precedent by the courts of many states, despite the fact it was decided almost 75 years ago.

It may be advisable at this point to state that this book differs materially from the typical case book used by law students. Such a book usually is composed of landmark cases. Such important cases, however, do not make their appearance in the law reports at regular monthly intervals, thus making it impossible to supply them to a magazine on a regular monthly schedule. Nonetheless, there is a fair share of landmark cases in this book.

The cases selected deal with many facets of construction work, and an effort was made to choose cases which include detailed descriptions of the facts involved so that the reader may put himself in the litigants' shoes and more readily understand how he might fare in a similar situation.

A lawsuit in the construction field often differs from the norm in an interesting and important respect. Instead of a clearcut struggle between a single plaintiff and a single defendant, the construction lawsuit is all too frequently a multi-party affair with all sorts of ramifications. Among the litigants may be found the owner, the contractor, the architect, the engineer, the supplier of materials or equipment, the insurance company, the subcontractor (often a bevy of them), the municipality, the injured workman, and the neighbor. Any one of that lot may turn up as a party in a construction action, and any one may sue or be sued by almost any one of the others. The result is a confusing situation that calls forth the highest legal skills and expertise of the judges charged with the unenviable task of resolving the controversies involved. If some of the decisions herein seem to contradict others, perhaps it is not the fault of the courts, but of the almost impossible task with which they are so often confronted in their efforts to do justice to all concerned.

Finally, readers of this book are cautioned not to regard these chronicles of controversy as giving a true picture of the United States' construction industry. The cases summarized herein are the tales of the things that went wrong with construction contracts and jobs, and it always should be borne in mind that for every disputed job that finds its way into the courts, there are a host of jobs that are completed to the satisfaction of all concerned.

An incident from the early days of my practice as a lawyer may help to prove that point. A contractor walked into the law office with which I was associated and said he needed an attorney because his blasting operations on a section of the New York subway system, which he was building, had caused neighboring property owners to sue him for damages to their buildings. In talking over the situation with him, we discovered that he had picked our office because he liked the names on the door and that he had previously satisfactorily completed a section of the subway without occasion to employ an attorney at any stage. Admittedly, he was fortunate, but his experience shows that it can be done, and it is done every day. If it were not so, the construction industry would be in a sorry plight.

It is hoped that this book and the cases recorded in it will prove not only interesting but also useful to engineers and others engaged in construction work, a factor so vital to the nation's economy and well-being.

William Jabine

Belfast, Maine

1. Plans and Specifications

The most important contribution to a construction project made by an engineer or an architect, or both, working either together or independently, is the preparation of the plans and specifications. For that reason the first three cases in this book deal with plans and specifications and decisions of the courts of various jurisdictions concerning their interpretations and functions, beginning with a case decided in 1899 and still frequently cited. It states the general principle that strict compliance with the plans and specifications is a requirement for the successful completion of a construction job. Subsequent cases herein incline to follow that principle.

How Accurate Must Plans and Specs Be?

A legal decision made in 1899, which still speaks with the voice of authority, clearly spells out the all-important role played by the plans and specifications in a construction contract and pinpoints the responsibilities and obligations imposed by them upon the parties to such contracts.

This case, over seventy years old, was decided by the New York Court of Appeals, that state's highest tribunal.

The MacKnight Flintic Stone Co. was the lowest bidder on a supplementary contract for the completion of a district courthouse and prison for the City of New York. The plans and specifications were prepared by the city's engineers.

By the terms of the contract, the plaintiff in the subsequent lawsuit, the MacKnight Flintic Co., agreed to furnish "all the materials and labor for the purpose and make tight the boiler room, coal room, etc., of the courthouse and prison . . . in the manner and under the conditions prescribed and set forth in the annexed specifications which are hereby made part of the contract."

The contract also provided for inspection by the city during the progress of the work, the inspector to see that the work corresponded to the specifications. As the floor of the boiler room was 26 feet below the curb level, and so probably below the tide level, there was danger of water pressure from below. Thus, the specifications went into complete detail in regard to the work to be done and the materials to be used in the waterproofing. Jackets for the columns were required and a waterproof lining over the floor and the lower walls were specified.

Although the contractor followed the specifications to the letter, leaks developed around the column jackets and in other places, and an ejector had to be installed to keep the boiler room dry. At this point the contractor proposed to do additional work to seal off the water, but his plan was rejected.

As the specifications contained a clause which read in part: "At the entire completion of the building all of the work must be gone over by the contractor and turned over to the city by him in perfect order and guaranteed absolutely water and damp proof for five years from the date of the acceptance of the work," the city refused payment for the work and the contractor brought suit. The plaintiff contended that it was impossible to obtain a dry cellar by following the specifications. After a trial the complaint was dismissed. The Appellate Division affirmed the dismissal and the contractor appealed to the Court of Appeals.

That court, in an illuminating opinion, excerpts from which appear below, reversed the ruling of the lower courts and held that the contractor, having followed the specification in every detail (except where he had used even more materials and done more work than called for), was entitled to a new trial at which a jury could pass on the facts.

It ruled that the defect was in the plan prepared by the city and not in the materials used, nor the workmanship. Under such circumstances, the court said, the contractor could not be held to be a guarantor of the sufficiency of the plan and so had done all that he was required to do. The court stated its reasons cogently and emphasized again and again the fact that the man who sticks to the specifications has done all that he has to do. These words are still being cited as precedents. The following quotations are from the opinion itself.

The rule of reasonable constructions governs courts in the enforcement of contracts. The contract now before us does not necessarily require the construction that the plaintiff guaranteed the sufficiency of the plan and specifications to produce the result desired, because it does not in terms so provide. There is no independent or absolute covenant to that effect. There is nothing in the subject of the contract, the situation of the parties or the language used by them, to conclusively indicate such an intention, and a fair and reasonable construction avoids such a peculiar and unjust result. The agreement is not simply to do a particular thing, *but to do it in a particular way and to use specified materials, in accordance with the defendant's design, which is the sole guide.* The promise is not to make water tight, but to make water tight by following the plan and specifications prepared by the defendant, from which *the plaintiff had no right to depart,* even if the departure would have produced a waterproof cellar. [Emphasis added]

If the contractor had designed and executed a plan of its own, which resulted in a tight cellar, it would not have been a performance of the contract, for it was to produce a waterproof cellar by following the plan and specifications made by the defendant and not otherwise. The plaintiff was not allowed to do additional work, according to a plan of its own, although it claimed it would prevent all dampness, and the defendant did not attempt to remedy defects at the expense of the plaintiff, as authorized by the contract.

There was no discretion as to the materials to be used or the manner in which the work should be done. The plaintiff had no alternative except to follow the plan under the direction of the defendant's officers in charge. The defendant relied upon the skill of its engineer in preparing the plan, with the most minute specifications, and bound the plaintiff to absolute conformity therewith . . . .

This is not the case of an independent workman, left to adopt his own method, but of one bound hand and foot to the plan of the defendant. The plaintiff had no right to alter the specifications, although the defendant had a qualified right to do so. If the plan and specifications were defective it was not the fault of the plaintiff, but of the defendant, for it caused them to be made and it alone had the power to alter them. It relied upon its own judgment in adopting them, not upon the judgment of the plaintiff. It decided for itself out of what materials and in what manner the floor should be constructed, and not only required the plaintiff to use precisely those materials and to do the work exactly in that manner, but also inspected both as the work advanced without complaint or question as to either . . . .

We think the evidence presented a question of fact for the jury as to the sufficiency of the plan to produce the result desired, and as to performance of the contract when properly construed. If the work was faithfully performed according to the plan and specifications, and the failure to secure a water tight boiler room was wholly owing to the defective design of the defendant, the plaintiff would be entitled to recover notwithstanding the refusal of the superintendent to give the required certificate, for under those circumstances it would be his duty to give it and a refusal to do so would be unreasonable. (*MacKnight Flintic Stone Co. vs. The Mayor*, 160 N.Y. 72; 54 N.E. 661)

If the Job Proves Difficult, Is the Contractor Liable for Performance?

No matter how "unusual" or "difficult of performance" plans and specifications may be, the contractor who undertakes the job for which they were written can be held to strict performance of them. His bid is submitted after ample opportunity to examine and study these specifications. If he unfortunately overestimates his ability to do the work in the manner prescribed, he must suffer the penalty therefor.

That is the essence of a decision made by the United States Court of Appeals for the Fifth Circuit, in a case concerned with the construction of a nuclear reactor facility in Montgomery County, Maryland.

The court noted that the novel character of the job itself was undoubtedly responsible for the complexity of the plans and specifications, but held that the subcontractor, whose work was in question, undertook said work with his eyes open and, therefore, should be held liable for his failure to meet the admittedly difficult requirements of the job.

The action was brought by the prime contractor, Blount Brothers Corp., against the insurance company which was the surety for the painting and special coatings subcontractor, William Dunbar Co., Inc., which, after many vicissitudes, had failed to complete the work it had undertaken. Blount based the action upon the claim that the sum required to complete the work after Dunbar's contract was terminated exceeded the balance due Dunbar.

The district court, after a jury trial, entered judgment for the defendant surety. The plaintiff, Blount, appealed to the Court of Appeals, Fifth District. As indicated above, that court reversed the judgment of the district court, holding that Dunbar had failed to comply with the requirements of the plans and specifications and that therefore the defendant surety was liable.

The opinion of the court of appeals outlined the situation as follows:

> On April 25, 1963, appellant, Blount Brothers Corp., entered into a contract with the United States of America, acting through the General Services Administration, for the construction of a nuclear reactor facility in Montgomery County, Md. This facility consisted of a nuclear reactor, located in a concrete containment building, and a supporting laboratory and office wing. On August 2, 1963, Blount entered into a subcontract with the William Dunbar Company, Inc., by the terms of which Dunbar agreed to perform certain portions of the prime contract, specifically,

the application of the special coatings in the nuclear reactor building and the painting in the laboratory and administration buildings.

The Standard Accident Insurance Co. was surety on Dunbar's bond. Standard was later merged with Reliance Insurance Co., which assumed all of the other company's obligations. By November 23, 1964, Blount, purporting to act pursuant to the authority of Art. XIII (a), of the general terms of the contract, notified Dunbar of a termination of the subcontract. Thereupon, Blount found another subcontractor who was engaged to complete the work (a substantial part) not already finished by Dunbar, and filed this suit against the bonding company to recover damages by reason of the fact that the subsequent contract cost substantially more than the unpaid balance due on the original contract with Dunbar. . . .

As might be expected from the fact that the work related to rooms to be used in connection with the housing of a nuclear reactor, the wall, floor and ceiling coverings, the subject of this contract, were of unusual (called by some "sophisticated") coverings, quite different from the ordinary interior house paint. Nevertheless, we must view the case in light of the fact that GSA desired to have this building completed exactly as prescribed in the specifications.

Blount Brothers Co. undertook to complete it in accordance with the same specifications, and Dunbar assumed the obligation to perform those parts of the job as fell within its contract to furnish the coverings. Of course, as required by the statute, the contractor and all of the subcontractors, including Dunbar, gave a performance bond to guarantee "all the undertakings, covenants, terms, conditions and agreements. . ." of their respective contracts. . . .

The historical facts relating to the status of the affair at the time of the letter of termination dated November 23, 1964, can be fairly simply stated. The subcontract required something unusual in the way of performance. It went much further than requiring that Dunbar cover the walls with a certain specified quality of wall coverings. It required, among other things, that, as to some of the work, the material to be used would have to be approved before application and it required, further, that Dunbar be approved by the manufacturer of the material as a qualified applicator, and that Dunbar also be approved by the GSA as an approved applicator.

These requirements applied to four of the thirteen types of materials that were to be used in satisfaction of the painting and covering subcontract. The types were numbered 1 to 13, but the materials required under categories 1 through 9 were promptly approved. The difficulty arose from the ultimate failure by Dunbar, at the time of termination to commence applying item 10 and to obtain approval of the unusual, but somewhat less sophisticated materials required for items 11, 12, and 13, and the approval of the particular materials to be used to satisfy these requirements.

The next few pages of the court's opinion contained a detailed chronicle of the various actions taken by Dunbar in its efforts to satisfy the strict requirements of the specifications.

The chronicle began by stating some of the difficulties as follows: "It required several pages of the specifications adequately to set forth the specifications for type 10. It required four other pages to specify the items which Dunbar was required to submit to the government in connection with obtaining approval of a proposed manufacturer of special coating type 10 . . . Types 11, 12 and 13 are non-elastomeric decontaminable coating intended for ceilings, walls and metal surfaces in the reactor area. It required four pages of the specifications to set forth the requirements as to these types, and six additional pages to specify the items which Dunbar was required to submit to the government in connection with obtaining approval of a proposed manufacturer of special coatings of types 11, 12 and 13."

The court continued by narrating in detail Dunbar's efforts to meet these exacting requirements. Dunbar was a franchise holder for one manufacturer and so attempted to have that manufacturer's product approved for the work. The product was rejected and notwithstanding an order to submit samples from another manufacturer, Dunbar resubmitted samples from the rejected manufacturer, an action which did not improve its standing with GSA.

Dunbar then got in touch with another manufacturer whose products were finally approved, but this manufacturer criticized the proposed methods of application. This precipitated new difficulties and brought forth a letter, described by the court as "What well may be considered an ultimatum" telling it to get on with the work. Several months of delay followed during which Dunbar sought to have various products and methods approved, including another resubmission of the first manufacturer. The termination of the contract by Blount followed.

After its thorough review of all the facts, the court concluded, as stated above, that Dunbar had presented no valid excuse for its failure to supply the required materials, and to perform its contract to apply them to the walls and other surfaces as set forth in the specifications. The court stated, in part:

> If, upon application of the proper legal principles, there was evidence which would have amounted to a defense against performing the requirements, this court could not review the jury's resolution of such factual issues. On the other hand, if there were substantial failures to perform, as was admitted, and there were no facts presented to the jury

which, under the law, would excuse performance, then the jury verdict should have been set aside on the motion, and the judgment should be reversed by us.

We conclude from a careful reading of the record, that as to several of the failures of the Dunbar Company, there were no issues of fact, which, if resolved in favor of Dunbar, would excuse such failure. We conclude, therefore, that as to the issue of breach of the contract, the verdict should have been directed by the trial court.

Appellee defends largely upon the concept that its bonded contractor had undertaken a very difficult subcontract which it was attempting, to the best of its ability, to carry out at the time of termination. A second defense is that at the time of termination others were at fault for the failure of Dunbar Company to satisfy the requirements imposed upon it by the subcontract.

With respect to the first defense, appellee's brief says:

"The fact is, and the record shows, it was Dunbar's contention throughout the proceeding (and the argument of counsel for appellee) that he did everything possible to obtain approval, exercised due diligence in doing so, and that the fault, and the force of the difficulty encountered in securing the approval of certain coatings, *lay in the specifications designed by the government.* [Emphasis added] This is what the record shows, and this is what the comment of counsel say."

It is difficult to see how the fact that Dunbar's difficulties arose from the specifications designed by the government could be in defense to Dunbar's failures. These were the specifications which the subcontractor agreed to be bound by. Moreover, the record is absolutely without dispute that it would have been extremely simple for Dunbar to have obtained approval of types 11, 12 and 13, if this subcontractor had been willing to approach sources of supply, the names of which were furnished to Dunbar by Blount early in the period of the contract. Instead of this, Dunbar, entirely possibly because it had an interest in selling Vortex paint and coverings, twice submitted this product after it had been rejected by the General Services Administration. . . .

The absolute failure of Dunbar to furnish any of the schedules or any of the lists of materials and names of suppliers of types 10, 11, 12 and 13, "before November 23rd," as required in Blount's letter cannot be excused by any testimony given on the trial, for, as above indicated, there was no proof that there was any bar to Dunbar's ascertainment of the names and qualifications of the suppliers of these types of coverings prior to that date.

The court then rejected Dunbar's contention that his failure to perform the work was due in part to defects in the surfaces to which the coverings were to be applied, and concluded by saying: "We find no evidence to warrant submission of the excuses for non–performance to the jury, because none of the evidence was sufficient as a defense to the requirements of the general terms of the contract and the specific terms of the subcontract."

The judgment of the trial court in favor of the surety was reversed. (*Blount Brothers Corp. vs. Reliance Insurance Co.*, 370 F. 2d 733.)

Does Deviation from the Plans and Specs Mean the Contractor Doesn't Get Paid?

The responsibility for a departure from the plans and specifications for a quarter–million dollar air conditioning and heating job, plus the fair penalty therefor, were the subjects discussed by the Supreme Court of Virginia in a lengthy opinion which ruled that the contractor had been guilty of an un-authorized deviation from the plans and specifications, but that the trial court had imposed an unfair penalty upon him by denying him the unpaid balance of the contract price for the work that he had done.

The contract called for the installation of air conditioning and heating systems in a building in Norfolk, at a contract price of $253,700. The job was to be completed by May 1, 1959. In its opinion the court described the subse-quent events as follows (the word *complainant* refers to the contractor, and the word *defendant* to the owner):

The contract, which was on a form approved by the American Institute of Architects, required the complainant to furnish all of the material and perform all of the work in accordance with drawings and specifications, made a part of the contract, which were prepared by Oliver & Smith, architects employed by the defendant. The architects and an independent engineer, James E. Hart, employed by them for this project, were, according to the contract, to have general supervision of the work.

The complainant presented evidence tending to show that as soon as the work commenced, it was discovered that the ducts, through which the air was to be carried from the conditioning unit to the upper floors of the building, could not be installed as prescribed in the drawings because of the presence of existing electrical conduits.

It was also asserted by the complainant that the room which was designated to house some of the necessary equipment was of inadequate size. The complainant contended, in the trial court, that these difficulties rendered the contract impossible of performance and necessitated changes in the work and the equipment to be installed, resulting in charges for extra work in excess of the contract price. On the other hand, the defendant presented evidence to show that the plans and specifications could have been complied with as originally prepared.

There was serious conflict in the evidence relating to the complainant's authority to make the changes in the work.

The complainant presented evidence, in the form of statements contained in a letter written by the complainant to the defendant after the work was accomplished to the effect that the defendant was informed of the necessity for the changes before they were made. However, the complainant's representative admitted on the witness stand that the defendant did not approve the changes, and the latter denied emphatically that he knew of any changes "until after the work had been completed, and when I had paid almost every cent under the contract."

The complainant attempted to prove that the changes in the work were authorized by Hart, the engineer, with the knowledge of the architects. However, it was shown that Hart had died after the work was completed but prior to the time of the hearings before the commissioner. Two letters, written by Hart to the architects after completion and which indicated his approval of the changes, were excluded by the commissioner. Oliver, the only one of the architects to testify, stated that neither he nor his partner, Smith, approved any changes in the equipment to be installed.

In any event, the work progressed to a conclusion, at which time the defendant had paid all but $14,473 of the contract price. On Dec. 22, 1959, the architect issued a final certificate attesting that the complainant was entitled to the payment of the balance due under the contract. The defendant was presented with the certificate, but he insisted that he be furnished with a "compliance certificate from the engineer and the architect and the materialmen who served the manufacturer" that the complainant "had delivered the tonnage" of air conditioning in the building. When the certificate was not forthcoming, the defendant refused to pay the balance of the contract price.

The complainant then filed a memorandum of mechanic's lien against the land of the defendant upon which the Law Building is located and later instituted this cause by filing a bill of complaint seeking a sale of the land to enforce the lien in the sum of $14,473 and a claim for extra work in the amount of $11,626.77.

The defendant filed his answer denying that he was indebted to the complainant and also filed a cross-bill praying for a judgment against the complainant for $75,000, alleged to be due because of the failure of performance and breach of the contract by the complainant.

The case was referred to a commissioner in chancery who heard the evidence of the parties. He filed his report in which he found that the air conditioning system installed by the complainant failed to meet the contract requirements because it was 46 to 50 tons "short in capacity"; the primary air unit was of a lower rating by 7,850 cfm; the condenser water pump was 120 gpm short of capacity; valves of inferior quality were installed; the size of water piping was reduced by 2 in.; pressure gages and balancing cocks were not provided on pumps; the duct insulation was inferior, and there were "many other items of equipment which did not meet mandatory requirements of specifications."

Based upon these findings, the commissioner ruled that the equipment installed was much less expensive and smaller in size and capacity than that called for by the specifications; that the defendant's consent was never obtained for any changes—major or minor; that there were substantial deviations from the contract requirements; that the complainant had knowingly departed from the plans and specifications so that there was no substantial compliance with the terms of the contract, and that the difference in cost between the equipment contracted for and that which was installed amounted to $24,252.50.

In his report, the commissioner recommended that the bill of complaint should be dismissed, "and that damages against the complainant should be awarded in favor of the defendant . . . in the amount of $24,252.50."

Both parties filed exceptions to the commissioner's report, which were overruled by the chancellor. A final decree was entered dismissing the bill of complaint and entering judgment on the cross–bill in favor of the defendant against the complainant in the amount recommended by the commissioner. The complainant was granted an appeal.

On this appeal, the complainant has not pressed its claim of $11,626.77 for extra work. It contends, however, that it was entitled to the enforcement of its mechanic's lien of $14,473 and that the defendant was entitled to recover nothing on his cross–bill.

The heart of the issue lies in the asserted authority of the complainant to make changes in the work. The complainant concedes that the changes were made and does not question that they were of a major nature.

In argument before us, counsel for the complainant stated, "The question is not whether there was a change in the plans and specifications—the question is, did the architect and engineer have the authority' to make the change?" The answer to that question is to be found in the provisions of the contract and the law applicable to a situation such as the one before us.

Before examining the wording of the provisions of the contract in the case before it, the appellate court stated the general rule as follows: "The contract between the complainant and the defendant provided that Oliver and

Smith, the architects, and Hart, the engineer, were to 'have general supervision of the work.' Under such circumstances, the architect is not, by virtue of his employment, the general agent of the owner for all purposes in the work he is engaged to supervise. His authority is a limited one—defined by the terms of his contract of employment or by the terms of the contract between the owner and the contractor. He has no authority to make alterations in the plans and specifications nor to bind the owner with respect thereto except as provided in the contractual documents. [Citations] "

After quoting from the provisions of the contract which flatly stated that written authority for changes, except in the case of minor changes not involving extra cost, must be obtained from the owner, architects or engineer, the court continued:

> Thus it is clear that the authority of the architects and the engineer to act as agents of the defendant was of limited scope, confined to those areas set forth in the contract and where, in special instances, their powers might be broadened. It is equally as clear that changes in the work, except of a minor nature, could only be made on the written order of the defendant or the written order of the architects or engineer stating that the defendant had authorized such changes.
>
> It was conclusively shown that the defendant did not, at any time, in writing or otherwise, authorize or approve the changes in the work. It was just as conclusively shown that neither the architects nor the engineer ever issued a written order stating that the defendant had authorized the changes. Nor was there any indication that the powers of these limited agents were, in any special instances, broadened so as to constitute them general agents of the defendant.
>
> There was, therefore, a complete lack of proof that the authority for the changes arose from the previously quoted provisions of the contract.

But the contractor had searched out other phrases in different parts of the contract which it contended gave the architects and engineer power to authorize changes without the written consent of the owner, including the architects' authority to stop work when they deemed it necessary, and a provision which gave them authority to "make decisions on all claims of the owner or contractor and on all other matters relating to the execution and progress of the work or to the interpretation of the Contract Documents." Bits of evidence tending to show that these powers had been exercised were pointed out.

The court, however, rejected the contractor's contentions, pointing out the inadequacy of the alleged proof and again emphasizing the necessity for written authority. On this point the court said in part:

But, whatever may have been said or done by the architects or engineer to lead the complainant to believe that they had approved the changes, their actions were in direct conflict with the provisions of the contract which required that written approval should be had before major changes could be made.

These contractual requirements were as well known to the complainant, and equally as binding upon it, as they were to the defendant and the architects and engineer. To the extent that the actions of the architects and engineer were in conflict with the provisions of the contract, such actions were in excess of the authority of these limited agents and, unless ratified by the defendant, not binding upon him.

The complainant says, however, that the defendant did ratify "the revision on the part of the engineer and architect" by remaining silent and not insisting "on the revision of the plans being referred to him for approval prior to the further execution of the contract. . . ."

. . . this contention flies in the face of the evidence, accepted and acted upon by the commissioner and the chancellor, that the defendant never authorized the complainant, the architects or the engineer to make any changes, that he never approved such changes and that he "did not know there was any change in the plans and specifications until after the work had been completed."

Knowledge is the essential ingredient upon which the theory of ratification rests. One cannot be held to have ratified that about which he has no knowledge. The record before us simply does not support the contention that the defendant knew about the changes while the work was in progress, and he certainly did not ratify them after he was given such essential information.

We must conclude that the complainant made the changes in the work without proper authority and that it deviated substantially from the requirements of the contract, to the damage of the defendant.

Although the Supreme Court of Appeals held that the complainant contractor had knowingly deviated from the plans and specifications, it did not agree with the lower court's ruling that this deviation should be penalized by depriving the contractor of the unpaid balance of the contract price.

After pointing out that the lower court had calculated the amount of damages by deducting from the value which the equipment would have had if the specifications had been followed, the value of the actual installation, the court discussed at length the cases cited and followed by the lower court in making its ruling that the contractor's conduct was such as to justify a forfeit of the balance of the contract price. It reached the following conclusion on this point:

It is clear from the record that the sum of $24,252.50, awarded on the cross-bill, was considered by the commissioner and the chancellor to be the outside amount of the damages caused by the changes in the work. The evidence in this case requires that the complainant be allowed a credit for the unpaid balance of the contract price against that figure. To hold otherwise would put the defendant in a better position than he would have occupied if the contract had been performed with precise exactness.

Furthermore, there is nothing in the record to suggest, and the commissioner did not so find, that the action of the complainant even approached the modicum of bad faith which would, in effect, visit upon it the imposition of a penalty for its departures. The record would indicate that, rather than setting out to save money on the contract by its wrongful betterment, the complainant spent more to perform the work than it was called upon to do.

The judgment was modified by deducting from the $24,252.50 awarded by the lower court the unpaid balance of $14,473, leaving the sum of $9,779.50 to be paid by the contractor to the owner (*Kirk Reid Co. vs. Fine,* 139 S.E. 2d 829.)

Can Plans and Specs Override Codes?

Carefully prepared plans and specifications, faithful adherence to them during the course of construction, and rigorous inspection during the progress of the work would seem to be the ingredients of a wholly satisfactory job. That is what a New York court, the Supreme Court, Dutchess County, determined when it was called upon to decide whether a plumbing contractor was subject to the regulations prescribed by a town plumbing board on a school construction job on which the court found that the above mentioned conditions were present. Under such conditions, which were prescribed by the state, the court held the plumbing contractor was not obliged to comply with the local regulations. The court stated the facts as follows:

This is an action to enjoin the defendants [called Hopper, Union Free School District and Central School District, respectively, in the order in which they are named as defendants herein] perpetually and during the pendency of this action. Plaintiff seeks this injunction to prevent the performance by Hopper of plumbing work in the construction of schools of Union Free School District and Central School District in the Town of Poughkeepsie until Hopper shall have submitted its plans and received a permit and certificate of competency from the Board of Plumbing Examiners of plaintiff and until it shall have employed a master plumber licensed by said Board. It sought further to enjoin the school district from contracting with or employing any person or firm who has not met the requirements of the plaintiff's Plumbing Board.

After stating the history of the case, including an appeal to the Appellate Division of the Supreme Court, which sent it back for trial under certain stipulations of fact, the court continued:

> From the pleadings and stipulation it is established that on November 2, 1964, Union Free School District awarded a plumbing contract to Hopper for the Hagantown Elementary School at a contract price of $102,250 and that on July 8, 1964, Central School District awarded a plumbing contract to Hopper for the Sheafe Road School at a contract price of $57,545. Hopper, a domestic corporation with principal offices at Yonkers, N.Y., was in the process of commencing its work on the Hagantown School and had performed a substantial part of its contract on the Sheafe Road School at the time it was enjoined.
>
> Plaintiff's Plumbing Code and Ordinance provides that all new plumbing work be inspected to insure compliance with requirements of the code and to assure that the installation and construction of the system is in accordance with approved plans (Article 13, Section 13.1.1); that it shall be unlawful for any person to work as a plumber unless he has passed an examination by the board of plumbing examiners and has received therefrom a certificate of competency and a license (Article 14, Section 14.1.1). A master plumber is one licensed to engage in the business of installing plumbing as a contractor (Article 14, Section 14.2.1) and to be licensed as such he must have such qualification as deemed necessary by the board of plumbing examiners (Article 14, Section 14.2.2).
>
> Upon the certification of the competency of an applicant and upon his registry and payment of a fee, a license shall be issued (Article 14, Section 14.2.5). Any firm or corporation engaged in the business of installing plumbing shall employ a master plumber licensed pursuant to the code (Article 14, Section 14.3.1). Before any plumbing work is commenced, a permit must be obtained by the master plumber (Article 14, Section 14.3.1). No permit shall be issued until detailed plans have been submitted assuring that they conform to the code provisions (Article 14, Section 14.3.2). Plans which indicate non-compliance with the code shall be rejected and no permit shall be issued until they have been revised (Article 14, Section 14.3.3).

After stating that Hopper had filed no plans and had filed no application for a license, and that the plaintiff admitted that the contracts had been entered into under the provisions of the General Municipal Law, the court proceeded to show how the provisions of that law safeguarded the work. It said:

The plans and specifications for the plumbing systems for each school were prepared by architects licensed to practice in New York State. They were approved by the respective school board of each defendant district. Pursuant to Section 408 of the Education Law, the plans and specifications were submitted to the New York State Commissioner of Education for his approval. The Department of School Buildings and Grounds, a division of the State Education Department, reviewed the plans and specifications to determine whether the plumbing system conformed to the state standards set by the Commissioner of Education. These standards appear in a building code compiled by the Commissioner for the construction of schools. The Department of School Buildings and Grounds, on behalf of the Commissioner of Education, approved the plumbing system for the Sheafe Road School on July 6, 1964, and approved the Hagantown plumbing system on October 30, 1964.

Upon said approval the contracts for each school were let to Hopper and work was begun by it. During the course of construction, inspections were made at each school. A clerk of the works was hired by each school board. Each was experienced in construction work with ability to supervise the installation of materials in accordance with the plans and specifications. They each had a minimum of 15 years experience as a supervisor on similar construction projects and each was hired by their respective school boards after an investigation and determination of their qualifications.

The clerk of the works was present on each job at all times and inspected all materials delivered to the job site to determine if the materials conformed to the specifications and if their installation was in accordance with the plans. Each clerk of the works made daily reports to the architect indicating daily progress, the results of tests conducted during the day and any defects in materials or workmanship.

The architect engaged by each school board is a member of the American Institute of Architects, licensed to practice in the State of New York and qualified by the Commissioner of Education to design plans and specifications for public school buildings. A field representative of the architect visits the job site at least three times per week to inspect the materials in the system to determine whether they had been installed according to plans. The architect visits the job site at least once every week to inspect the materials and system.

A firm of plumbing engineers, licensed to practice in the State of

New York, was retained by the architect to inspect and test the plumbing system to determine that the installation is in accord with the plans and specifications.

Upon completion of the work the architect is to make a final inspection and test on all plumbing work and thereupon will issue his certificate to each school board that the plumbing work is complete and that the system conforms to the plans and specifications as approved by the Commissioner of Education.

The plans and specifications provide for tests for the drainage system, water lines and mains and gas lines, all in the presence of the architect. The plumbing work completed at the time of the injunction was inspected and tested and complied with the Building Code of the Commissioner of Education and the plans and specifications approved by him.

Upon these facts, the issues are well defined: Is the plumbing contractor who performs this work within the Town of Poughkeepsie in accordance with the requirements of the State Education Law and the General Municipal Law and the plans and specifications for the erection of new public schools under contract with the respective school boards subject to the regulations of the plaintiff's Plumbing Code?

Two defenses are raised: First, that Hopper is exempt from the provision of the code because it is working on public school contracts and second, that it is unconstitutional. Since this case may be disposed of by reason of Hopper's exemption from the provisions of the code, question as to the constitutionality of the code is not reached.

The court then considered a number of cases cited by the parties including one that seemed to hold that the local regulations were valid. In regard to this case the court said, "The court is aware of the decisions of *City of Kingston vs. Bank,* 45 Misc. 2d 176, 256 N.Y.S. 2d 276, enjoining an unlicensed plumber from performing a contract for the Kingston Board of Education in violation of the city plumbing ordinance. Special Term in granting the temporary injunction herein relied on this decision. In that case, however, the plans and specifications were submitted by the architect engaged by the board of education to the city plumbing inspector and the specifications themselves provided that inspections were to be made by the city plumbing inspector as well as by the architect. The defendant applied for a certificate of competency, was examined and failed. It thus may be argued that the voluntary submission to the control and regulation of the municipal plumbing code in the Bank case distinguishes it from the case at bar."

The court then ruled that a school board is immune from a local plumbing or building code and again emphasized the thoroughness of the State's regulations. It said in part:

> Although plaintiff's code provides for approval of plans and specifications and the inspection of work performed thereunder, plaintiff now concedes that it does not have the power to pass upon plans and specifications. It must take this position because of the provisions in the Education Law providing for the approval of plans and specifications by the Commissioner of Education (Sec. 408, Education Law). Provisions for compliance with the Commissioner of Education's regulations to insure the health and safety of the pupils in relation to proper heating, lighting, ventilation, sanitation and health, fire and accident protection (Sec. 409, Education Law) and the regulations adopted by the Commissioner of Education setting forth the requirements to obtain the approval of the Commissioner of Education for plans and specifications submitted to him. (Regulations of the Commissioner of Education. Subchapter J, Part 155, Section 155.1 *et seq.*)

> These provisions show clearly that the legislature has preempted the area of plans and specifications and vested it exclusively in the Commissioner of Education. Plaintiff urges that beyond this point, i.e., in the field of actual construction, there has been no preemption. It claims that the legislature has not acted in this sphere and that to insure the health and safety of the pupils and the community, the police power of the town must be exercised in the enforcement of its planning code and ordinance.

> Plaintiffs cannot sustain this contention. The legislature clearly has given the Commissioner of Education and the school boards the power and obligation to see that schools are properly constructed. . . . The proof herein shows the elaborate and adequate inspection and supervision of every stage of a construction by qualified persons to insure the erection of the buildings in accordance with the plans and specifications, the adequacy of which has not and cannot be questioned.

> Having conceded that it has no control over the plans and specifications, plaintiff cannot enforce the balance of the code against Hopper. Permits are issued only when the plans comply with the code provisions (Article 14, Section 14.3.2) and inspections are made to insure compliance with the code and to assure installation and construction of the system in accordance with approved plans (Article 13, Section 13.1.1).

> Since the plans and specifications are those of the school boards approved by the Commissioner of Education and not subject to review by the plaintiff's plumbing board, the issuance of a permit and the in-

spection as provided by the ordinance become meaningless and can have no applicability to Hopper. (*Town of Poughkeepsie vs. Hopper Plumbing & Heating Corp.*, 260 N.Y.S. 2d 901.)

Can Bidders Rewrite Ambiguous Specs?

A California contractor brought an action against the County of Stanislaus contending that the County had required him to do work on a housing project that was outside the limits shown on the architectural plot plan. He demanded extra payment for this work. Although it was evident that the drawings were ambiguous, the trial court ruled against him, and he appealed. The Fifth District Court of Appeal of California upheld the ruling of the trial court and in so doing, discussed thoroughly the roles of the various constituents of a construction contract and their relations to each other. Following is what the court said.

Plaintiffs–contractors appeal from an adverse portion of a judgment in declaratory relief construing their duties and obligations under a housing development construction contract. Defendant, Housing Authority of Stanislaus County, called for bids for the construction of housing units at two sites within the county. Plaintiffs were the successful bidders and entered into a single contract with defendant covering both projects. Although details differ for the work involved at each project, only one legal question is presented: Under the contract are plaintiffs required to install sewer and drainage lines and manholes beyond the project limits? The trial court found that plaintiffs must make all plumbing installations shown on the drawings or mentioned in the specifications, necessary to make the projects usable.

The question arises because part of the sewer and drainage lines, connections and manholes shown on the drawings and necessary to connect the dwelling units to existing main lines, extend beyond the "contract limits," "property line," and "project limits" shown on the architectural plan. Plaintiffs contend the drawings delimit their responsibility to the designated boundaries. However, the lines and manholes are clearly shown on the drawings, and the plans indicate where existing 6 in. lines are to be replaced by 8 in. pipe. Furthermore, the fall or grade of the drainage line is indicated on the drawings, and a detail of the "off–site" storm drain anchor block is shown.

The drawings, standing alone, are ambiguous since they not only designate the limit lines upon which the plaintiffs rely, but also designate sewer lines and installations necessary to complete each unit, that extend beyond the limits. However, the drawings constitute only one document in an integrated contract that provides:

"The contract shall consist of the following component parts: (a) This Instrument, (b) General Conditions, (c) Special Conditions, (d) Technical Specifications, and (e) Drawings."

In determining the meaning of a contract all documents that are parts thereof must be construed together. [Citations] Hence the drawings must be interpreted in the light of the other documents incorporated in the agreement.

To eliminate uncertainty in case of a possible discrepancy between the drawings and technical specifications, the general conditions provide that:

"Anything mentioned in the Technical Specifications and not shown on the Drawings, or shown on the Drawings and not mentioned in the Technical Specifications, shall be of like effect as if shown or mentioned in both. *In case of difference between Drawings and Technical Specifications, the Technical Specifications shall govern.*" [Emphasis added]

The technical specifications, which prevail over the drawings, require that the contractor shall:

"b. Provide all materials and appurtenances necessary for the complete installation of each utility, whether or not all such materials and appurtenances are shown on the drawings or described in the specifications."

The trial judge also noted that by the first paragraph of the contract plaintiffs agreed to furnish

"All labor, material, equipment and services, and to perform and complete all work required for the construction of Low Rent Housing Projects CAL.26.6B, Patterson, consisting of Ten (10) dwelling units, and CAL.26-8, Westley, consisting of Twenty (20) dwelling units, together with necessary site development including but not limited to grading; paving; sewers. . ."

This provision the trial judge interpreted as contemplating usable dwelling units which, of course, entailed sewers connected to the mains and storm drains connected to sewage facilities:

Special conditions in the specifications delineate work to be done by others at no expense to the contractor. The exempted work, including the electrical distribution system to and including meters and street lighting systems, but not off-site plumbing, sewer or drainage facilities. Failure to include "off-site" plumbing in this itemization of facilities the contractor is not required to install means, conversely, that he is required to furnish them. He said:

Plaintiffs argue if this court should conclude that the contract requires them to make the disputed plumbing installations, they are entitled to reimbursement on the theory of unilateral mistake. They cite *M. F. Kemper Const. Co. vs. City of Los Angeles,* 37 Cal. 2d 696,235 P.2d 7, but in that case the city had knowledge the contractor had inadvertently omitted a $301,769 item from a $780,305 bid *before* the bid was accepted.

In the instant case, plaintiffs' plumbing subcontractor testified that in order to submit the lowest bid possible he intentionally confined his bid to the limits shown on the drawings. He noted these limitations and restrictions on the bid he submitted to plaintiffs, thus giving notice to plaintiffs prior to the time they submitted their bid. Unlike the *Kemper* case, these facts were not made known to defendant until after the bid had been let, the contract was signed, and construction was well under way. As pointed out by the trial judge, "Instructions to Bidders" provide that if a bidder is uncertain as to the interpretation of the specifications and drawings, he could secure verification of interpretation before bidding. The facts of this case are not opposite to those of *Kemper.*

Based upon the quoted provisions of the various documents comprising the complete contract, the appellate court, as stated above, ruled in favor of the defendant county, and against the plaintiff contractor. It thus put the stamp of its approval on the great care with which the contract had been drawn with its provision for the very sort of contingency which occurred, a variation between drawings and specifications. It also emphasized the fact that a prospective bidder on a construction job must, for his own protection, examine all of the documents which make up the contract with great care *before* he submits his bid. A discovery made after the bids are opened and the contract awarded will come too late, even though the fault may to some extent be mutual, or may even be that of the owner. (*Meyers vs. Housing Authority of Stanislaus County,* 50 Cal. Reptr. 856.)

Does Performance Make the Contract Binding?

Both sides of an interesting question involving construction contracts were set forth in the prevailing memorandum and the dissenting opinion in a case decided by the Appellate Division of the Supreme Court of New York, Second Department. The following quotations will explain the situation:

(*From the prevailing memorandum*) On Nov. 14, 1962, plaintiff and the corporate defendant Oxford signed a contract for the installation of the central cooling and heating plant in defendant's building, in accordance with certain plans and specifications, all work to be done in accordance with plaintiff's letter of Oct. 26, 1961.

The contract contained an arbitration clause. It also provided that the contract was not to become effective until the plans and specifications were signed by the contracting parties.

Thereafter plaintiff completed the job except for certain work which was not completed at defendant's request. When plaintiff commenced this action defendants moved to compel arbitration, which plaintiffs opposed on the ground that the plans and specifications had never been signed. It is conceded that plaintiff completed the work in accordance with the plans and specifications. Under these circumstances, both contracting parties have by their conduct adopted the plans and specifications, the failure to sign them is inconsequential, there was compliance with a valid agreement (CPLR 7503), and plaintiff should be compelled to arbitrate. [Citations]

(*From the dissenting opinion filed by two justices*) To hold, as does the majority, that completion of the work constitutes a waiver of the provision that the contract should not become effective until both contracting parties signed the architect's plans and specifications, is to extend the contract, and necessarily, the arbitration provision thereof, which may not be done. [Citation] If a party wishes to bind another in writing to an agreement to arbitrate future disputes, this purpose should be accomplished in such a way that each party to the arrangement will fully and clearly comprehend that the agreement to arbitrate exists and binds the parties thereto. [Citation]

Here the plain meaning of the condition precedent is that the contract and its arbitration provision do not effectively exist and do not bind the parties unless and until they both have signed the architect's plans and specifications. Consequently if plaintiff's completion of the work amounted to a waiver at all, it was a waiver of the written agreement *in toto,* and not just in part. (*Race Co. vs. Oxford Hall Contracting Corp.,* 268 N.Y.S. 2d 175.)

2. Exceptions to Strict Compliance

The cases in this section indicate that the courts do not always insist on the classic strict compliance with the plans and specifications, but are willing to consider what might be termed *extenuating circumstances* and other, similar factors. The importance of specifications and plans is not belittled, but neither are they deemed immutable.

What Is Substantial Compliance?

In the course of determining the rights of the parties under two "companion" contracts for the construction of a motel, the Supreme Court of Oklahoma set forth its interpretation of the law concerning two matters of importance to all engaged in the carrying out such construction contracts.

One of these rulings dealt with interference by the owner resulting in deviation from the plans and specifications with unfortunate consequences. The other thoroughly discussed the important question, "What constitutes such substantial compliance with the plans and specifications that the contractor is entitled to recover despite a certain measure of deviation from said plans and specifications?"

The two contracts were entered into at the same time by the owner of the lots on which the motel was to be built and a firm of contractors. The first was for construction of the motel according to the plans and specifications for the sum of $83,000. The second provided that the owner would sell to the contractor a one-fourth interest in the completed motel.

As is so often the case, things did not go as smoothly as the parties to the contracts had anticipated when they were signed. The owner, Alice M. Collins, brought an action against the contractor, D. D. Baldwin and the

Baldwin Construction Co., alleging breach of the construction contract. The contractor sought to enforce his rights under the contract for acquisition of a quarter interest in the motel.

After a long trial and the compilation of a record of more than 1,000 pages of testimony, the District Court of Oklahoma County ruled that the contracting firm had substantially performed the construction contract even though it had failed to comply with certain of the specifications, such as repair of the swimming pool and certain phases of the air conditioning equipment. The damages to the plaintiff caused by these deviations were determined by the court to be $15,484.50.

This amount was to be credited to the plaintiff on the purchase contract but she was denied rescission of the purchase contract for which she had asked. Arrangements were made for placing the motel in receivership until the rather complicated situation could be resolved and the interests of the opposing parties could be reduced to dollars and cents.

The plaintiff, however, was not content with this solution of the problem and appealed the judgment of the trial court to the Supreme Court of Oklahoma. The high court affirmed the judgment of the trial court ruling that there had been substantial performance of the construction contract—in spite of the $15,000 worth of deviations—and that the contractor was entitled to enforce his right to purchase a one–fourth interest in the motel after the necessary adjustments had been made. In a nine–page opinion, the court discussed the two questions of interest previously referred to.

There were several complaints by the contractor charging interference on the part of the owner. One was set forth in considerable detail by the court as follows:

> Before commencement of the motel's construction, the surface of the lots, comprising its planned site, had an elevation at the back, or eastern side, that was much higher than their front, or western side, facing Lincoln Blvd. At its highest point, the site's eastern boundary was as much as 12 ft higher than said street.
>
> As the plans for the motel required the eastern wall to be built only 12 in. west of the lot's eastern boundary, with only a gradual slant in the paved automobile parking area and driveway in front of the motel and extending to the Lincoln Blvd. curb line, it was necessary to excavate as much as 6 or 7 ft below the east side of the lots to lay the concrete footing and/or foundation on which the motel was constructed.
>
> This resulted in its ground floor being several feet below the surface of the ground east of it along the lot's eastern boundary, and made the motel's eastern wall similar to a retaining wall, in that most of the masonry was below that adjacent ground level; and, between said surface and the ceiling of the motel's first story, there was little more than

enough room for the air conditioners that were installed in holes left therefor in each room's east wall, with about half of the thickness of the conditioners protruding outside the wall. The fact that the surface of this ground east of the air conditioners was in some places higher than the bottom edge of these air conditioner holes in the motel's east wall created a problem which appears, from the evidence, to have been recognized by the builder during the motel's planning stages.

According to the evidence introduced on his behalf, it was satisfactorily solved by modification of the structure's plans and changing the "grade" along its east wall, so that the surface water, which might otherwise have drained toward that wall, was diverted both to the north and to the south, and would drain toward the streets on either side of the motel. There is no question from the evidence but that within a few weeks after the plaintiff went into possession of, and started operating, the motel, she caused changes to be made in the "grade," or surface of the ground immediately east of the motel's eastern wall, which defendant's evidence tended to show permitted surface water to seep into that wall at the level of the air conditioner openings.

There was also evidence tending to show that this water seeped out of the wall onto the floors of some of the rooms about baseboard level, and/or higher. Some of the plaintiff's evidence, in effect, conceded that this could have accounted for some, or all, of the damages plaintiff claimed to the carpeting of said rooms from surface water as aforesaid. Here, again, a question of fact was presented.

We think that, on the basis of the evidence, the trial court would have been warranted in finding that the leakage of the surface water into these rooms was due to none of the claimed defects in the motel's construction, but rather to the plaintiff's change, after the motel's completion (and as already indicated) of the drainage pattern that had previously been arranged by M.D. and/or D.D. Baldwin, acting for the defendant's construction company.

The supreme court's discussion of what constitutes *substantial* compliance with the plans and specifications occurs in that portion of its opinion in which it disposes of the plaintiff's contention that the trial court's ruling that there had been substantial compliance and its award of more than $15,000 in damages to her were contradictory. The court said, in part, quoting from *Kizziar vs. Dollar* (10th Cir.) 268 F.2d 914:

. . . The law is settled in Oklahoma that when a contractor and builder has in good faith endeavored to comply with the terms of a contract, literal compliance in all details is not essential to recovery, especially where the owner has taken possession of the building. In

Robinson vs. Beatty, 75 Okla.69,181 P. 941, 942, the Oklahoma Supreme Court said:

"Since the rule of exact or literal performance has been relaxed, literal compliance with a building contract is not *essential to a recovery thereon, but a performance thereof in its material and substantial particulars is sufficient.* [Emphasis added]

"There is substantial performance when the builder has in good faith intended to perform his part of the contract and has done so in the sense that the building is substantially what is provided for, and there are no omissions or deviations from the general plan which cannot be remedied without difficulty. . . ."

In *Bushboom vs. Smith,* 199 Okl. 688,191 P. 2d 198, we said: ". . . In such case the judgment should make such monetary award to the injured party as would place him in the position he would have been in had the contract been performed, but it should not put him in a better position than he would have been had there been complete performance."

In its application of the rule here dealt with, the court in *Kizziar vs. Dollar, supra,* appears to have considered significant the fact that the building involved was constructed for use as a medical clinic and that it had been successfully so used for a considerable period. Likewise, we think that in this case it is significant that the defects complained of by plaintiff in the building involved have apparently constituted no insurmountable, or even substantial, obstacle to her acceptance and use of it for the purpose for which it was constructed and intended.

In her argument, plaintiff does not directly question the good faith of the builder in his efforts to comply with Contract 1, except to charge him with ignoring his duty to build the motel in a "workmanlike" manner, and of being guilty of unspecified "unscrupulous acts." From our examination of the record it is our opinion that the builder's good faith in such endeavor cannot be successfully challenged.

In view of this state of the record, and, as none of the defects have prevented the premises from being occupied as a motel, and it has been so operated with some degree of success, we think the requirements of the above rule have been met and that the prerequisites for the operation and efficacy of Contract 2 came into being. In other words, while the subject motel's business may have been hurt or damaged, to some extent by such things as a defective filter system in the swimming pool, the failure of its air conditioners to both heat and cool its rooms satisfactorily, a squeaking floor in an upstairs room, and/or by some claimed defects in construction, for which the trial court awarded no damages, none of such defects were of such a serious character as to prevent, or drastically curtail, its use in accord with the objects of Contract 2, nor have they kept plaintiff from enjoying the benefits contemplated therein.

To cancel that contract and allow her to appropriate to her own exclusive advantage the benefits derived from defendant's work and investment in his performance of Contract 1, to the substantial extent to which it was performed, would result in her exclusive ownership of a business property, which the evidence seems to indicate was worth considerably more when its operation was begun than the consideration she gave for it; and, contrary to the above quoted admonition in *Bushboom vs. Smith, supra,* would put her in a better position now than she would have occupied had defendant completely performed all of his obligations "to the letter" in Contract 1. (*Collins vs. Baldwin,* 495 P. 2d 74.)

It is doubtful some of the more conservative courts would go as far as the Oklahoma Supreme Court does in its relaxation of the old rule that specifications must be strictly complied with, but this case is of importance because of the weight it attributes to the good faith of the builder and the fitness of the structure for the use for which it was intended.

Which Comes First—Specs or Public Interest?

Are specifications that limit the bidders to one designated product compatible with the concept of competitive bidding on public construction projects? That was a question the Court of Appeals of Ohio, Montgomery County, was called upon to answer.

Montgomery County proposed to build two municipal incinerator plants. Plans and specifications for the two structures were prepared and distributed among prospective bidders. Before the date for submission of bids, a taxpayer's suit was brought seeking a writ of *mandamus* that would require the county commissioners to disregard all bids on the ground that the specifications were illegal because they called for the prime contractor to use a designated subcontractor for a major portion of the work.

The court's first task was to decide if the action had been properly brought because of the fact that the petitioner who sought to have the specifications declared invalid was an incinerator subcontractor who desired to submit a bid but who was barred from doing so by the limitation to a designated subcontractor supplying a designated product. The county maintained that the petitioners sought pecuniary benefits for themselves rather than for the benefit of the public. On this point the court said:

A taxpayer's action *in mandamus* may be maintained by a party in a private capacity to enforce the right of the public to the performance of a public duty, as distinguished from a purely private right of the taxpayer to the performance of a duty imposed upon a public servant.

State ex rel. *Nimon vs. Village of Springdale,* 6 Ohio St. 1, 215 N.E. 2d 592.

At page 4, 215 N.E. 2d at page 595 the court said that: "In a long line of cases, this court has repeatedly recognized the rule as stated in 35 Ohio Jurisprudence 2d 426, Section 141, that 'where the question is one of public right and the object of the *mandamus* is to procure the enforcement of a public duty, the people are regarded as the real party and the relator need not show that he has any . . . special interest in the result, since it is sufficient that he is interested as a citizen or tax-payer in having the laws executed and the duty in question en-forced. . . .' "

By analogy it follows that the existence of and "special interest in the result" should not defeat the right of a taxpayer to maintain such an action.

Inasmuch as the petitioning incinerator subcontractor was a taxpayer in Montgomery County, the court upheld its right to maintain the action.

The court then turned to the main issue of whether or not the specifica-tions should be disregarded because they limited the choice of the incinerator subcontractor to a single designated firm. It began by quoting a number of significant provisions from the specifications. Some typical provisions follow:

Qualifications of Incinerator Subcontractor.

He shall have been, and shall be currently exclusively engaged in the furnishing and installing of municipal incinerators for the past fifteen (15) years.

He shall have furnished and installed equipment of the type he proposes in at least five (5) municipal plants in different counties or cities in the United States. The five plants must currently be operating and giving satisfactory service.

Each drying grate shall be a type specifically designed for the pre-drying of municipal refuse in an incinerator and shall be manufactured and installed by the Incinerator Subcontractor or the principal Owners. The Incinerator Subcontractor or the principal Owners shall have manufactured this particular grate for this service for at least twenty (20) years and he shall have manufactured and installed drying grates for at least five proven municipal or county installations now in service using the same type grate of which at least one installation has been in constant satisfactory service for at least twenty (20) years and at least two additional installations which have been in constant satisfactory service for at least five (5) years.

Similar specifications calculated to bar newcomers to the business, regardless of their competence, and to ban new products, no matter their merits, were written for ignition grates and rotary kilns.

This portion of the specifications was climaxed by the following statements:

> The furnishing and installing of the incinerator and complementary equipment is a part of the contract and the cost of same is to be a part of and included in the amount bid.
>
> This equipment will be furnished and installed by:
>
> International Incinerators, Inc., Walton Building, Atlanta 3, Georgia, Phone—Jackson 3-1678.
>
> All bidders qualified as General Contract Bidders may obtain a copy of the complete incinerator proposal from the above company. A copy of this proposal is on file with the Engineers. General Contract Bidders may confirm their proposals at this location.

The court turned next to the fundamental question of whether or not the provisions it had quoted from the specifications met with the statutes' requirements for competitive bidding on public projects. It said in part:

> The respondent board was required by Section 153.50, Revised Code, to seek from general or "prime" contractors, separate proposals for furnishing materials and work necessary to installing incinerating equipment in the buildings to be erected. The board, in its discretion, could invite proposals for materials and work separately or for both combined.
>
> The respondent board chose, in the instant case, to do neither. Rather, it invited bids under specifications which limited each general or prime contractor bidder's discretion as to the subcontractor whose services or material he would use to a single firm, i.e., International Incinerators, Inc.
>
> Such action is clearly an effort to avoid the salutary provisions of the statute requiring competitive bidding on public contracts, whatever hair–splitting arguments may be made concerning some of the concededly ambiguous provisions in Chapter 153, Revised Code. We conclude, accordingly, that the specifications hereinbefore described, more particularly that in paragraph 19, designating International Incinerators, Inc., as the only acceptable subcontractor for the manufacture and installation of incinerator facilities in the structures in question, are not in conformity with the statutes requiring competitive bidding on county contracts.
>
> It has been said that no chain is stronger than its weakest link. The incinerator machinery, its complementary equipment and its installation

are a part, and probably the largest and most important part, of the whole contract. Respondent claims the right to so write its specifications so as to exempt this part of the contract from competitive bidding. The mandatory provisions of the statutes requiring competitive bidding on the whole contract must be held to apply as well to its component parts relating to each separate trade or kind of mechanical labor, employment or business and for the furnishing of materials therefor. Any other interpretation of these statutes would rob them of their intended force and effect. . . .

We recognize the right of respondent to establish "Qualifications of Incinerator Subcontractor" in their specifications . . . . We fully appreciate the benefits to the public, which would follow the award of the work to a firm with the skill and experience and a product that would meet the respondent's specifications. But there is the law to be taken into account. The Legislature must have believed that in the long run the public interest would best be served by the enactment of statutes requiring competitive bidding on public improvements. The Supreme Court has expressed the same opinion by making strict adherence thereto mandatory.

Considering the specifications relating to Incinerator Subcontractors together with the Instructions to Bidders set forth in the petition, we must conclude that respondent's specifications have effectively stifled competitive bidding for the incinerators, their complementary equipment and their installation. It may be that evidence would show that these matters should be excepted from competitive bidding. But, we are required to render our judgment upon the facts alleged by relators and admitted by respondent's demurrer. We consider that one who seeks to excuse himself from the operation of mandatory statutes should be required to prove the facts warranting such exception. . . .

We conclude that the facts alleged by relators and admitted by respondent show that respondent is about to enter into a contract which is illegal because it is made without compliance with the statutes requiring competitive bidding. Such facts further show that relators have capacity to maintain this action as taxpayers and that, in the absence of any evidence of record showing the product and services are absolutely unique, relators have a clear legal right to the relief sought. (*State vs. Board of County Commissioners of Montgomery County*, 229 N.E. 2d 88.)

Is Slavish Following of Specifications Enough?

It has become a fairly common practice to include in certain types of construction contracts a provision that promises the completed facility will produce certain stated results.

In some cases, such as contracts for the installation of heating or cooling systems, a stated temperature is promised, and in others, such as in the case discussed below, a given number of units to be produced is promised. Thorough consideration was given to such a contract provision or specification in this case by the United States District Court, Eastern District of Tennessee, which was called upon to decide a breach of contract dispute between a tile manufacturing company and a designing and construction company which had designed and built a new kiln for the tile company.

The suit was brought in a federal court because the tile manufacturing company was in bankruptcy and it was necessary for the trustee in bankruptcy appointed by the federal court to bring the action on behalf of the plaintiff.

The court stated the situation as follows:

The plaintiff, Hood Ceramic Corp., formerly known as the B. Miflin Hood Co., was for many years engaged in the production and sale of quarry tile, along with related products, with its plant located at Daisy, Tenn.

Quarry tile is a form of tile used in construction, particularly in tile floor construction, it being most often seen in red and buff color. It appears that Hood was at one time perhaps the largest producer of this type of tile in the nation.

For many years, in fact until 1958, the plaintiff fired its tile in periodic kilns, described as "beehive" kilns by reason of their appearance. In more recent years a continuous kiln, known as a "tunnel" kiln, also by reason of its appearance, had been developed in the industry and installed by many of Hood's competitors so that by 1949 Hood found itself in a poor competitive position and rapidly losing its dominance in the market.

It therefore undertook negotiations with the defendant for the design and installation of a tunnel kiln, which negotiations ultimately and after lengthy delays, due in part to the ill health of the president of Hood and in part to Hood's difficulty in financing construction of a new kiln, led to the execution under date of May 7, 1957, of the contract herein sued upon.

The defendant, Ferro Corporation, through its Allied Engineering Division, with offices located in Cleveland, Ohio, is engaged in the business of designing and constructing kilns for the production of quarry tile. It appears to bear a most favorable reputation in the field of designing and constructing tunnel kilns. As noted above, negotiations between the parties extended over a period of years, but ultimately resulted in the execution of a contract under date of May 7, 1957. By the terms of this contract the defendant was to design and construct a tunnel kiln for Hood at its plant in Daisy, Tenn., for the price of

$190,000. Attached to and made part of the contract was a set of general specifications relating to the kiln. . . .

(At this point the court inserted several of these specifications, including the following: "Estimated production–5,000 sq ft 6 in. x 6 in. per 24 hours. Estimated total firing cycle–78 hours. Estimated schedule–2 hours and 10 minutes per car.")

The court continued by stating that the construction of the kiln was completed in January, 1958; that some of Ferro's men remained to supervise the starting of the kiln and made needed adjustments, and did not leave until December of 1958; that Ferro was paid $190,000 from the proceeds of a loan made to Hood by the United States Small Business Administration; and that in August, 1961, a voluntary petition in bankruptcy was filed by Hood.

The court defined the issues before it as follows:

There are three major issues raised in the trial of this case which require decision by the court. The first issue involves the construction of the contract entered into by the parties for the design and construction of the kiln.

The plaintiff contends that the defendant contracted to design and erect a tunnel kiln that should have a firing cycle of 78 hours, would operate on a schedule of one kiln car each two hours and ten minutes, and fire a minimum of 5,000 sq ft of 6 in. x 6 in. x ¾ in. quarry tile or a minimum of 7,500 sq ft of 6 in. x 6 in. x ½ in. quarry tile in a 24 hour period, seven days a week and 365 days a year, and that the tile thus produced would be not less than 95 percent standard or marketable grade.

In arriving at this construction of the contract, the plaintiff further contends that the contract is ambiguous and that parol evidence of prior negotiations is admissible in arriving at a construction of the contract. The defendant upon the other hand contends that its obligation under the contract was only to properly design a kiln along the lines described in the specifications, in which the production schedule and quantity were only estimates, and then erect a kiln in accordance with that design.

The defendant denies that it specified that it would construct a kiln that would produce any particular percentage of marketable or standard grade tile and contends that no ambiguity exists in the contract in this respect which would render admissible parol evidence to vary or add to the specifications in this regard, but rather that paragraph X of the contract specifically provides that the written agreement constituted the whole agreement between the parties. The defendant further denies that the parol evidence relied upon by the plaintiff would in fact establish any specification with regard to the production of a minimum percentage of marketable or standard grade tile.

The second major issue for decision is as to whether the evidence establishes a breach of contract by the defendant. In this regard the plaintiff has introduced evidence which it contends establishes that only 41.25 percent of the tile fired through the kiln from January 1, 1958, through August 21, 1961, the date of bankruptcy, was of standard or marketable grade, with the kiln, by reason of defective design and construction, causing cracking or other defects in the remaining tile fired through it.

The defendant, upon the other hand, contends that no breach of the contract has been shown, that the evidence establishes that after the initial adjustment of the kiln in operation, the kiln operated satisfactorily, produced up to 80 percent to 95 percent standard grade tile, with cracking and defects in the tile being caused not by the design and operation of the kiln, but rather by the plaintiff's other antiquated equipment and methods and by numerous other production faults and errors.

The third issue related to the measure and proof of damages; Hood contended that it was entitled to the sum of $11,000, which it had expended in bringing the kiln up to specifications and for its loss of profits on the sale of tile because of the kiln's failure to perform as promised. The defendant denied all liability on these grounds.

Then the court took up the principal issue, the question of whether or not the contract should be interpreted as providing that a specific quantity of marketable tile would be produced within a specified time. On this issue the court said:

Returning to the issue of the construction of the contract, the court is of the opinion that the specifications are ambiguous with regard to the production to be designed and constructed into the kiln and that parol evidence upon this issue was properly admitted. [Citation] The contract specifies, "estimated production—5,000 sq ft 6 in. x 6 in. per 24 hours." Only two dimensions of the tile are given in this specification, the thickness of the tile being omitted. The thickness of the tile is significant in that approximately one-third or more tile of ½ in. thickness could be loaded upon each kiln car than would be the case of tile ¾ in. in thickness.

The evidence reflected that a tile ¾ in. in thickness was produced by Hood prior to the installing of the tunnel kiln and that Dr. Robson, who was vice president of Ferro and manager of the Allied Engineering Division and was in charge of the design of this kiln, had visited and inspected the plaintiff's plant upon several occasions prior to drawing the contract and specifications and was aware that Hood produced ¾ in. tile. Although some contention appears to have been made on behalf of Ferro that tile ½ in. thick was more common in the industry, it was conceded by Dr. Robson that the reference in the specifications to 6 x 6 tile might appropriately be interpreted as referring to either or both tile ¾ in. or ½ in. in thickness. The court is therefore of the opinion that in specifying estimated production the contract had reference to 5,000 sq ft of 6 in. x 6 in. x ¾ in. per 24 hours. The quantity of tile which the kiln was to be designed to fire in any 24 hour period was therefore 5,000 sq ft of 6 in. x 6 in. x ¾ in. tile or 7,500 sq ft of 6 in. x 6 in. x ½ in. tile.

While the quantity of tile that may be fired in a given period is thus specified, no reference is made in the contract or in the specifications to any percentage of marketable tile that may be fired. The only purpose of the kiln was to produce marketable tile. Otherwise it was worse than valueless, for it could only be a source of financial loss. The argument on behalf of defendant that the defendant's only obligation was to construct a kiln of "proper" design appears specious unto the court until the term "proper" is related to the production of usable or marketable tile. Surely no valid argument could be advanced that a kiln was properly designed if it would not produce usable ware, regardless of the quantity that might be fired. The correctness of its design and construction can ultimately be tested only by the results produced in terms of usable tile. Surely no valid argument could be made that a kiln which could produce no usable or marketable tile was properly designed and constructed.

The critical factor to be designed and built into a continuous kiln is not just to raise the heat sufficient to fire the tile and then cool the tile, but to do so in such a regulated, even and controlled manner that usable tile will be produced. Thus it appears unto the court that omission of any reference to usable or marketable or standard grade tile to be produced by the kiln would create an ambiguity in the specifications that may properly be supplied by parol evidence.

The parol evidence which was admitted in conformity with the court's finding of ambiguity in the wording of the written contract as set forth above, consisted in part of a letter written by Dr. Robson of the Ferro organization, in the course of the contract negotiations, which read, ". . . You will produce 5,000 sq ft of tile per day, seven days a week, 365 days per year.

"You will produce an average of not less than 95 percent standard or saleable tile. Actually our latest figures from one plant are 97 percent to 98 percent first grade ware. This ware is on uniform color and absorption so that necessary adjustments due to faulty tile in the job are entirely eliminated."

After quoting this letter, the court said, "The court is of the opinion that the negotiations between the parties were conducted with the understanding that the kiln would be designed so as to produce 95 percent standard grade or marketable tile and that this evidence may be looked to in clarification of the specifications."

Although the court ruled in Hood's favor so far as interpretation of the specifications were concerned, it also ruled that the evidence in regard to loss of profits due to the inadequacy of the new kiln did not establish such loss with a sufficient degree of certainty to permit a finding for Hood on this score. Accordingly, the court allowed Hood to recover the sum of $11,000 in recompense for the money paid to put the new kiln in proper working condition, but, on the other hand allowed Ferro the sum of $29,247.78 plus interest on its counterclaim, which included an indebtedness to it of $4,247.78 on open account and $25,000 due upon a note. (*Clark vs. Ferro Corp.*, 237 F. Supp. 230.)

Can Specs Be Unreasonable?

The United States Court of Claims softened the rigidity of the "or other equally suitable material" provision that is a part of the usual construction contract when it held that, no matter what the specification might seem to say, the rule of reasonableness must be applied in determining the real meaning of the contract. The court overruled the decision of various government boards that had denied a contractor payment for materials he contended were equal in in quality to those actually described in the specifications.

The use of the word *rigidity* in the foregoing paragraph is especially suitable because the specifications in the contract before the court, a contract for additional air conditioning facilities at Fort Belvoir, Va., required a rigid type of insulation for the protection of certain air ducts while the contractor proposed to supply a blanket type of insulation. The government engineers insisted on the rigid type and the contractor brought suit to collect the extra $3,000 it had cost to furnish the rigid insulation. The court stated the facts of the situation as follows:

Plaintiff's claim arises from the alleged wrongful refusal of the contracting officer to approve, for use in the insulation of supply and return ducts, not exposed within air conditioning spaces, blanket (flexible) material submitted by the contractor as represented to be equally suitable material to the block or board (rigid) type of insulation named in certain specifications, which were incorporated in the technical provisions of the contract by reference. Following the aforestated action of the contracting officer, plaintiff installed the more costly board type insulation on the ducts in question. Thereafter, plaintiff filed a claim for additional compensation which was denied by the contracting officer on January 24, 1961.

There followed a chronicle of the contractor's successive, unsuccessful appeals to various government boards culminating in the action before the court. The court's opinion continued:

> The case is here on an Assignment of Errors filed by plaintiff attacking the administrative decisions of both the Engineers Board and the ASBCA [Armed Service Board of Contract Appeals] as being arbitrary, capricious, and erroneous as a matter of law. A responding statement was filed by defendant [the United States] to which plaintiff submitted a reply. . . . The sole issue presented to the court is the question of liability.

The court then quoted in full the provisions of the specifications in question, placing special emphasis on the following words that appeared in the specification describing the insulation material, ". . . or other equally suitable material approved by the contracting officer."

The court continued:

> It is apparent from examination of sections 3.6 and 3.6.1 of the above Federal Specifications that the contract specified block or board type insulation for concealed ducts. Plaintiff does not dispute this fact, but it alleges, on behalf of itself and its said subcontractor, that in the preparation of its bid, all concealed duct insulation was estimated on the basis of using blanket type insulation, and that such estimate was justified by reason of an established trade usage permitting the use of blanket insulation in the performance of contracts such as the one involved here. Plaintiff attempts to fortify the aforestated position by pointing to the language of the technical provisions of the contract (Section 25-32 [e] *supra*), which allow the use of "other equally suitable [insulation] material."
>
> The facts essential to a decision are not in serious dispute, and for our purposes here, the evidence established by the testimony presented to the Engineers Board is adequately summarized in that Board's opinion and decision of March 7, 1962. It appears therefrom that about 1951, when a satisfactory fiberglass blanket material appeared on the market, contractors throughout the country began offering the use of blanket type insulation on non-exposed ducts during the course of performing government construction contracts covering work similar to that required under the contract involved in the instant case. This was done for each job by the contractor requesting approval of the government for a sub-

stitution of materials on the ground that the blanket insulation was superior, rather than by attempting to show compliance with the appropriate federal specifications designating block or board type insulation.

The evidence before the Engineers Board disclosed that the Corps of Engineers entered into a number of construction contracts which contained a provision identical, or substantially similar, to Section 25-32e of the Technical Provisions, *supra,* and incorporated by reference Federal Specification HH-I-562, *supra.* Many of the contracts were administered by the Washington District Office of the Corps of Engineers involved in this appeal, and even some by the same area office under said District. Testimony was presented that in all instances the blanket insulation was accepted by the Corps of Engineers without any negotiation of a credit to the government.

In connection with the foregoing, it is considered significant to note that the record shows plaintiff's subcontractor, i.e., the General Insulation Co., had performed the insulation work on approximately 12 prior contracts in which the blanket insulation had always been approved and accepted by the same Corps of Engineers District Office involved in this appeal.

The court quoted at length from the decisions of various government boards that had denied the plaintiff's claim for additional compensation. In every case, the decisions clung to the theory that, despite the fact that blanket insulation might do a better job and had been widely used for almost a decade, the Contract Officer's decision that it was not "equally suitable" was final and could not be disturbed. One decision even hinted that the government officials who had been accepting blanket type insulation might have been derelict in their duties.

The court did not agree with these arguments. Concerning dereliction on the part of officials who had approved blanket insulation, it said, "Such an assumption is not warranted by the record. If assumptions should be indulged in at all, it seems more logical to assume that authorized government personnel felt justified in approving the blanket insulation on the authority of the terminology here in dispute. And certainly no authority need be cited for the proposition that prior administrative interpretations are entitled to great weight."

The court continued, "It is clear plaintiff did not consider that the specifications furnished by the government were unclear, ambiguous, or subject to a construction different from the one given to them by plaintiff. The record shows that plaintiff prepared its bid, and planned and attempted to perform the contract work, in a manner consistent with its interpretation and construction of the contract specifications. Plaintiff gave a meaning to the specifications which, in the light of its knowledge of how similar contracts had been treated in the past, was not unreasonable or in any way improper under the circumstances here."

The defendant, the United States, also put great stress on the contention that the provision of the contract that was in question was unambiguous and capable of only one interpretation—that made by the various boards of appeal. The court disagreed:

> In my opinion, the technical provisions and specifications incorporated in the contract are ambiguous, and this fact is illustrated by the conflicting interpretations placed upon them by the parties, which joined in giving birth to this lawsuit. When these provisions are considered together, as they must be, it is apparent that more than one reasonable conclusion can be reached. That is to say, it appears to me that the meaning of these provisions is not so explicit as to put the contractor on notice that only rigid insulation would be accepted. On the contrary, the phrase "other equally suitable material" could very well be meaningless, if that construction were adopted. The contention of the defendant bears this out for it does not offer any explanation of how these provisions can be harmonized, short of ignoring the job specification. Had the word "rigid" or an equivalent been inserted in the contract so that the specification read "other equally suitable *rigid* material," the case would take on a different appearance. In this attitude of the controversy, however, I am unable to conclude, as a matter of law, that these provisions are subject to only one reasonable conclusion. . . .
>
> What has been decided at this juncture of the discussion herein is that a flexible insulation, if "equally suitable," complies with the contract specifications. If the government desired to assure that only rigid material would be installed, it should have taken the necessary steps to add to, or change, the job specifications. Since the clause "or other equally suitable material" was included *in toto* without any expressed qualification, it must be accorded that meaning which is reasonable and abiding in the trade.
>
> It is apparent that the basis of the Engineers Board's refusal to consider evidence of trade usage, custom or practice was its holding that the contract was unambiguous. Thus, my conclusion that the contract was ambiguous crumbles the props upon which the decision below rests. Evidence of trade usage, custom and practice was properly before the Boards and should have been considered. But even in the absence of any ambiguity, the Board committed legal error by failing to consider such evidence. . . .
>
> As noted previously, the contract and the Wunderlich statute vested discretion in the contracting officer to approve or disapprove of insulating material. It is apparent from what has already been said that

he was not justified in disapproving of blanket insulation. The contracting officer disapproved the substitution of blanket material, not because it was inferior but because he interpreted the contract as giving the government the right to insist on rigid insulation, even though the blanket material was equal in all respects. What the government did here was to mislead plaintiff into bidding upon the basis that an "equally suitable material" could be used and then, through the device of fact finding, refused to find any but the rigid material satisfactory. As noted earlier, this is a determination which is not unassailable in this court for it is a legal issue.

While the government had the undisputed right to decide if the blanket material met the standards set forth in the contract documents, "it is equally elementary that the discretion involved must be exercised reasonably and fairly." [Citations] The government's arbitrary and capricious interpretation of the specifications, and refusal to grant an equitable adjustment for the change so brought about, were grossly erroneous actions as a matter of law. That is sufficient to vitiate its adverse decision on the plaintiff's claim. However, conceding for the sake of argument that a question of fact is being reviewed, the same result is reached because the record shows that the contracting officer did not consider evidence of trade usage, custom and practice. Thus, the decision that only rigid type insulation complied with the specifications was erroneous, as it lacked in content the necessary ingredient of support by substantial evidence. Accordingly, an equitable adjustment should have been granted under these circumstances. [Citation]

Those who wish to read the full text of this notable victory of flexibility over rigidity, both on the job and in the courtroom, will find it in the nearest law library. (*W. G. Cornell Co. of Washington, D. C. vs. United States,* 376 F. 2d, 299.)

What Happens When a Contractor Deviates from Specs?

The Court of Civil Appeals of Texas decided a lawsuit in which the court was compelled to explore in detail the principles governing deviations from plans and specifications, as well as the rules defining the penalties for deviations. The County of Tarrant brought suit against the contractor for a courthouse and jail, against the architects who designed the structure, and the engineer (a member of the architectural firm) who prepared the plans and specifications. Although the county claimed damages of $128,721.95, the jury awarded it $700 against the architect, who in the ensuing appeal did not contest that nominal award. Here, then, is what happened.

In 1959, the Commissioners' Court of Tarrant County (Fort Worth) decided to erect a building for a courthouse and jail. It employed as architects Birch D. Easterwood & Partner, who agreed to do the work for less than the usual fee because they were relieved of the obligation of regularly inspecting the construction. They prepared the plans and specifications. Butcher & Sweeney Construction Co. agreed to construct the building according to those plans and specifications. R. S. Smith, a structural engineer who was a member of the architectural firm, prepared the part of the plans and specifications dealing with structural requirements.

The county sued Butcher & Sweeney for damages caused by a breach of its contract because it deviated from the plans and specifications in these ways:

It attached shelf angles to the concrete wall with studs, not with the expansion bolts that the plans and specifications called for; it installed shelf angles with 3-in. legs projecting horizontally from the concrete wall, not the 5-in. legs called for, and it failed to install brick anchors 24 in. on centers, vertically and horizontally.

A jury agreed that Butcher & Sweeney deviated from the plans and specifications in those ways. But as for the construction company bearing the cost of remedying the defects, which the county placed at $128,721.95, the jury answered, no. It also found that the defects caused by the deviations could not have been remedied after the building was completed without unreasonable economic waste and that the deviations did not reduce the value of the building.

The construction contract authorized the architect to approve minor changes not involving extra cost and not inconsistent with the purpose of the building. The jury found that substituting power-driven studs for expansion bolts in attaching shelf angles to the concrete wall was that kind of minor change; that Easterwood had authorized Smith, the engineer, to act for him in permitting the substitution; that E. W. Thomas, acting architect for Easterwood, had approved that substitution; that installing shelf angles with 3-in. instead of 5-in. legs projecting horizontally from the wall was a minor change not involving extra cost and not inconsistent with the purpose of the building; that Easterwood had likewise authorized Smith to act for him in permitting installation of shelf angles with 3-in. legs; that Smith had authorized installation of 3-in. legs, and that it was the practice in Tarrant County at the time for a structural engineer, such as Smith, who was associated with an architectural firm, to act for such an architect in approving minor structural changes.

And as to any knowledge the Commissioners' Court of Tarrant County may have had of such deviations, the jury found that before the building was finished, the county's project inspector and the county engineer knew of the deviations and that one commissioner knew during construction that power-driven studs were being used in attaching the shelf angles to the concrete wall. The jury further found that the county commissioners, in the exercise of ordinary care, should have known of the deviations during construction. But the representatives of Tarrant County, according to the jury, failed to complain to the construction company of such deviations with reasonable promptness after the county's project inspector and engineer learned of them. The jury found that this failure by the county, however, was not a proximate cause of the defects caused by not installing through-wall flashing and weep holes at the base of the penthouse.

The county sued architect Easterwood for damages caused by his negligence in failing to provide in the plans and specifications for expansion joints in the brick veneer walls and through-wall flashing and weep holes at the base of the penthouse walls. The county also sued Smith, who they alleged was negligent because he authorized Butcher & Sweeney to attach shelf angles to the concrete walls with nails instead of expansion bolts. The county sued Easterwood and Smith for the reasonable cost of remedying the defects caused by their alleged negligence.

The jury found that Easterwood failed to use the degree of care that an architect of ordinary competence and skill in Tarrant County would have used in the same or similar circumstances by omitting from the plans and specifications requirements for through-wall flashing and weep holes at the base of the penthouse walls and that it was this failure that was a proximate cause of defects in the building. It found the county was damaged $700 by that failure, and the court rendered judgment for Tarrant County against Easterwood for that amount. The jury did not find Smith guilty of any act of negligence that was a proximate cause of damage. It did not find that Butcher & Sweeney breached its contract or did any unauthorized act or that the county was damaged by its deviations from the plans and specifications. The county appealed these two parts of the judgment.

Butcher & Sweeney presented evidence that its substitutions were of equal materials and methods, if not better; that they were duly authorized; and that they were known by the county's project inspector during construction. It did not disagree that expenditures were required by the county to repair damage to the building that occurred after the county accepted it. It was apparently agreed that the sole cause of damages, other than that assessed against Easterwood, was the differential movement between the inner concrete wall and the brick veneer wall, for which the county acknowledged that the contractor was not responsible. Under the jury findings, those deviations were authorized and did not cause damage. The appellate court said:

> The county's first "proposition under many points of error," is that since the pleadings and evidence established that the building was substantially complete, the proper measure of damage was the reasonable cost of remedying the defects. It says the question is whether Butcher & Sweeney failed to perform its contract; that it is undisputed

that it did fail; that the county paid the contractor for something it didn't get, a building completed according to said plans and specifications. It says that, because the jury found Butcher & Sweeney deviated from the plans and specifications, it is liable to the county for the cost of remedying the resulting defects. As shown, the jury found that the cost of remedying the effect of said deviations was "none." Appellant says such finding is contrary to all the evidence or, in the alternative, so contrary to the overwhelming weight and preponderance of the evidence as to be manifestly wrong. All Butcher & Sweeney's established deviations from the plans and specifications were, according to the findings of the jury, authorized, were approved and were known to the county's project inspector and caused no damage. We are forced to the conclusion that the answers complained of by appellant are supported by evidence of probative force and that they are not contrary to the overwhelming weight and preponderance of the evidence.

Appellant further says the trial court erred in giving effect to the finding that the deviations found could not have been remedied after completion of the building without unreasonable economic waste. It says the correct measure of damages to be assessed against Butcher & Sweeney Construction Co. for failure to comply with the contract was the reasonable cost of remedying the effect of its deviations and therefore the court erred in giving effect to the finding.

[Appellant] says the question of economic waste was not involved and the phrase "without unreasonable economic waste" had no proper part in determining the measure of damages; that economic waste is material only where there is not substantial performance of the contract, in which event the measure of damages is the difference in the value of the building as constructed and what the value would have been if it had been completed according to plans and specifications. It says that an exception to the rule that such difference in value is the proper measure of damages is applicable when there has been substantial performance by the contractor. . . .

Appellant also says the court erred in submitting the issue that inquired what reduction in value was caused by such deviations, to which the jury answered "none," and that the court erred in giving effect to that answer because there was not submitted an ultimate issue of fact. It says the current measure of damages is the reasonable cost of remedying the defects caused by the failure of Butcher & Sweeney to comply with the plans and specifications. We think the argument must be rejected. The jury found that no damage was caused by Butcher & Sweeney's deviations from the plans and specifications and, further, that they were authorized by the construction contract. Butcher & Sweeney say the jury has properly found that to remedy the effect of deviations from the plans and specifications, excluding the damages caused by differential movement, for which the contractor was admittedly not responsible, involved economic waste; that the evidence shows the contractor used as good or a better method than that prescribed by the plans and specifications and that the undisputed evidence is that the building was safe and that no repairs were required as a result of the deviations found and that, under the facts and findings relative to economic waste, the proper measure of damages was the difference in value of the building as constructed and what its value would have been if it had been constructed according to the plans and specifications.

The court then cited this rule, previously laid down:

In whatever way the issue arises, the generally approved standards for measuring the owner's loss from defects in the work are two: First, in cases where the defect is one that can be repaired or cured without undue expense, so as to make the building conform to the agreed plan, then the owner receives such amount as he has reasonably expended, or will reasonably have to spend, to remedy the defect. Second, if, on the other hand, the defect in material or construction is one that cannot be remedied without an expenditure for reconstruction disproportionate to the end to be obtained, or without endangering unduly other parts of the building, then the damages will be measured not by the cost of remedying the defect, but by the difference between the value of the building as it is and what it would have been worth if it had been built in conformity with the contract.

The court continued:

Butcher & Sweeney say that economic waste may be considered when repairs are required to make a building safe. We think this is correct if such unsafe condition is attributable to deviations from the plans and specifications. If the building was unsafe for other reasons, that condition cannot be attributed to the Butcher & Sweeney's deviations from the plans and specifications. The evidence shows the building was safe, except as it may have been made otherwise by the county adopting plans and specifications which contained no provision for the contingency of differential movement. Butcher & Sweeney agreed only to construct the building according to said plans and specifications. According to jury findings, it did construct the building in accord therewith and, further, no damage resulted from the deviations found by the jury.

Under the findings that Butcher & Sweeney's deviations were

authorized and that they caused no damages, appellant's contention that the correct measure of damages is the reasonable cost of remedying the defects caused by the deviations of Butcher & Sweeney from the plans and specifications cannot be sustained. . . . The county is not entitled to recover damages against Butcher & Sweeney for breach of contract under the jury findings that the deviations were authorized and no damages resulted. . . .

The jury found that Butcher & Sweeney deviated from the plans and specifications in three particulars. But this was not a finding that it breached its contract. The jury also found that said deviations constituted minor changes, not involving extra cost and not inconsistent with the purposes of the building. Such changes were authorized by the contract. . . .

The jury also failed to find that engineer Smith, who with Butcher & Sweeney constitute the appellees here, was guilty of any act of negligence which damaged the county. The county had the burden of establishing a breach of the contract by Butcher & Sweeney and that it was damaged thereby and, as to Smith, that he was guilty of some act of negligence alleged and that it was a proximate cause of damage to the county. There being no such findings, the county was not entitled to judgment against appellees.

We have considered all of appellant's contentions, ably presented in a lengthy brief. All appellant's points are overruled. We conclude that reversible error is not shown. The judgment is affirmed. (*County of Tarrant vs. Butcher & Sweeney Construction Co.*, 443 S.W. 2d 302.)

Is There a Difference Between Substantial and Complete Performance?

What constitutes a *substantial* performance of a construction contract compared with a *complete* performance was discussed in an opinion handed down by the United States Court of Appeals for the Eighth Circuit.

Fenestra Inc. (subcontractor) sued Nathan Construction Co. (primary contractor) to recover $49,817.84, which Fenestra claimed Nathan owed it under a 1962 construction contract for windows and a curtain wall for an Omaha building. Fenestra alleged full performance on its part.

Nathan asserted that the curtain wall was of inferior design and workmanship. Nathan contended the wall did not comply with the specifications because it permitted excessive air, water, and dust infiltration; did not move freely with changing temperatures; created excessive stresses and distortions; was improperly sealed; and subject to abnormal deterioration. Nathan filed a counterclaim for $395,000 in damages based on its alleged inability to fulfill its obligations to Julius Novak, the building's owner; or, in the alternative, for indemnification against any liability on its part to Novak. The owner was permitted to intervene, under Civil Rule 24 (b), as having a claim with questions of law and fact in common with Nathan.

The case went to trial without a jury before Chief Justice Richard E. Robinson of the Nebraska District Court. He found that Fenestra had demonstrated substantial performance of its obligations under the subcontract, that with substantial performance shown, Nathan and Novak had the burden of proving any damages they had incurred, and that they had not sustained this burden and were not entitled to any setoff. Accordingly, judgment was entered in favor of Fenestra for the $49,817.84 balance, and Nathan's counterclaim and Novak's claim were dismissed. Nathan and Novak appealed to the Eighth Circuit Court.

The building in question was at 3002 Farnam St. in Omaha. Novak contemplated remodeling it into a luxury apartment house–office building. Nathan's agreement with Novak contained the usual provisions for compliance with specifications and, in particular, called for the installation of a curtain wall. By the subcontract between Nathan and Fenestra, the latter agreed to furnish and install windows and the curtain wall. The stated consideration was $105,000. This contract contained provisions relating to workmanship, infiltration and thermal expansion.

On the one hand, Nathan and Novak contended that Fenestra failed to prove performance or even substantial performance according to the specifications; that even if substantial performance was shown, Fenestra did not prove the cost of remedying the difference between substantial performance and complete performance; and that the burden of proof of that cost was on Fenestra.

Fenestra, on the other hand, argued that if it had fully performed its contract, it was entitled to the contract price; that if it had only substantially performed, it was entitled to the contract price reduced by any damage sustained by Nathan and Novak; that the district court's findings that Fenestra had shown substantial compliance were amply supported by the evidence; that the burden of proving damage was Nathan and Novak's; and that they had failed to sustain that burden.

The appeals court then noted with emphasis this narrative of events:

The curtain wall was not anything new in Omaha. Curtain walls had been used in other renovation work in the area. The exterior design was by Novak's architect and was determined before Fenestra acquired its status as subcontractor. Fenestra prepared shop drawings, and adjustments in these were made in accordance with conditions encountered on the site. Fenestra's superintendent on the job worked with the contractor's superintendent there.

Fenestra's work was open and obvious. It was never told to stop its work. Representatives of the contractor and Novak himself inspected the project frequently.

Fenestra regarded its work as complete by June, 1963. It inspected the curtain wall and detected no infiltration of air or weather.

Nathan, however, refused to pay the balance due on the subcontract. Representatives of Fenestra, Nathan, and Novak met several times to ascertain the basis of the complaints and what might be done to remedy them.

Another meeting was held in Omaha in January, 1964. It was attended by Novak and representatives of Nathan and Fenestra. The building was inspected; no evidence of leakage was found.

They met again in April. Fenestra had a weather-stripping man install an aluminum sample on vents. Novak watched the installation and spoke favorably of it.

Fenestra ordered enough of this material to weather-strip the vents. Fenestra was instructed to install it in vacant units first. These then constituted about 25 percent of the building. When the weather-stripping was completed in the unoccupied apartments and in one which was occupied, a Nathan representative ordered Fenestra to stop work and remove its crews from the building. Novak stated that a reason for this was that there was a continuous hum from the stripping. There is some dispute whether the installed stripping was ever removed. Shortly after the work was stopped and during a heavy wind and rain storm, Fenestra representatives inspected the building and vents. Again they found no water or air infiltration.

Still other meetings took place in July, August, and September of 1964. They produced no specific demonstration to Fenestra that the complaints of water and air infiltration were supported.

At a May, 1966, meeting, a complaint was made that caulking had failed. Fenestra personnel were unable to find evidence of caulking failure but did note that some neophrene was loose. They also found some cracking in surface paint, which they attributed to the caulking's elastic ability to "move" with the grid system; paint does not do this. There also was a complaint that snow had come in around one vent in an unoccupied apartment. Inspection disclosed that the vent was open, but it was unknown how it came to be open.

A final meeting took place in November, 1966, almost three and one-half years after Fenestra's work was completed. There was one sagging neophrene member and a few loose ones on the outside of the building. A complaint was made that light was coming in through the curtain wall. Fenestra representatives contended the light came up into the ceiling plenum from the room below through space between the ceiling and the curtain wall. A card was inserted and pulled through the space. Fenestra had not installed the ceiling.

The curtain wall is so constructed that it is free to move by expansion and contraction as temperature varies. The wall's components are affixed by tape and caulking, and, on the exterior, by a neophrene glazing bead. All permit movement. Each grid unit is independently attached to the building by welding. "There was testimony," the court noted, "that there was no evidence of stress on the welds and that the grid system is self-draining by the channeling of water into the vertical members and down to the bottom of the wall."

The owner's architect acknowledged that the specifications recognized some permissible infiltration in the ventilators and that infiltration depends on wind velocity and pressure.

Some apartments in the building were first rented in July, 1963. The rest were ready about November. Since then, every unit has been rented at some time. The vacancy rate for the 119 units has been as low as six percent. Rentals run from $135 to $425 a month. Novak himself described the building at one time as "plumb full."

After stating that Nebraska law applied and that the burden of proof that it had complied with its obligations was on Fenestra, the appeals court made these points:

A. *The intent of the parties.* The defendant and the intervenor argue that weather protection was important for this reconstruction job and was deemed necessary and highly desired by them. There is little question as to this. . . . However, we do not read the subcontract as one that required that the building be *absolutely* water-and-weathertight, with nothing due Fenestra if that high standard was not fully met. Indeed, despite prior references in the agreement to nonallowance of "any weather infiltration" and to gasketing "to prevent infiltration of weather," the quoted Paragraph 14 recognizes the possibility and permissibility of some infiltration: "Air infiltration of ventilators shall not exceed 1/3 cfm/ft of crack when closed." What is imposed is a limitation, not an absolute guarantee of no infiltration.

Our issue then, as the defense points out in its brief, is whether plaintiff constructed the type of weathertight curtain wall called for by the specifications.

B. *Fenestra's posture.* Plaintiff Fenestra at all times has taken the position not that it rendered only substantial performance of its subcontract with Nathan but that it fully performed its obligations under that agreement and that it is therefore entitled to recover the balance of the stated contract price. The complaint alleges, "That plaintiff has fully and properly performed its said subcontract with defendant and is entitled to be paid in full therefor." Judge Robinson observed this and said, in his memorandum, "Plaintiff's position in its complaint and throughout the trial was that it had completely performed its obligation under its subcontract with Nathan Construction." Thus it was the trial court which advanced the concept of substantial performance into the decision of the case and made findings in respect thereto.

C. *Substantial performance.* It has been said that American courts are united in holding that a substantial performance of a building or construction contract will support a recovery either on the contract, or in some jurisdictions on a *quantum meruit* basis. Three reasons for this have been advanced: 1. that work on a building is such that, even if the owner rejects it, he receives the benefit of it and it is equitable to require him to pay for what he gets; 2. literal compliance with every detail of specifications is impossible; and 3. the parties are assumed impliedly to have agreed to do what is reasonable under all the circumstances with respect to performance.

Substantial performance, of course, is not easily defined. Fortunately for us, in this diversity case, the Nebraska Supreme Court has announced its definition of the concept and has recognized substantial performance in litigation of building and construction contracts.

"While it is difficult to state what the term 'substantial performance' or 'substantial compliance' as applied to building and construction contracts means, it seems that there is substantial performance of such a contract where all of the essentials necessary to the full accomplishment of the purposes for which the thing contracted for are performed with such an approximation to complete performance that the owner obtains substantially what is called for by the contract." (*Jones vs. Elliot,* 174 Neb. 96, 108 N.W. 2d 742, 748 [1961].)

The appellate court pointed out that Judge Robinson, using that definition, found that Fenestra had shown substantial compliance and that the owner had received substantially what he bargained for.

"We are in full accord with this holding," the court said, "and find it most adequately supported in the record. . . ."

The court appended at this point an eleven-point list of the ways Fenestra had fulfilled its obligations. It then continued:

> From the entire record we gain the impression that Fenestra in good faith leaned over backwards in a sincere effort to ascertain the grounds of the Nathan and Novak complaints; that, in its many meetings and inspections, in its willingness to weatherstrip, in its reinforcements of already installed brackets, and in other ways, Fenestra did all which could reasonably be expected of it to give consideration to the complaints and to remedy any which were justified; that its efforts thus to satisfy were not a concession of inadequacy of performance; that, actually, the record is supportive of a finding of complete, rather than only a substantial, performance; and that there was substantial compliance under any analysis of the facts. Were we the trier of facts, we would be inclined to conclude that Nathan and Novak received all they bargained for and that Fenestra had demonstrated complete performance.

Of course, it is always possible that the work could have been better performed. This is so with respect to almost any kind of work including, we might note, judicial decisions.

We therefore reject the argument of the defendant and intervenor that substantial performance of the subcontract was not shown.

After ruling in favor of Fenestra on the main issue of the case, the court devoted a couple of pages to Nathan's $395,000 counterclaim. It ruled that even if such a legal claim should exist, the defendant had failed to prove the amount so expended and so could recover nothing.

The judgment of the trial court in favor of Fenestra for the entire amount demanded, $49,817.84, was affirmed. (*Nathan Construction Co. vs. Fenestra Inc.*, 409 F 2d 134.)

3. Interpreting Plans, Specifications, and Contracts

The five cases in this section are pertinent to the interpretation of plans and specifications and of the contracts of which they are integral parts. The third case is concerned with the interesting question of what happens in the fortunately rare situation when plans and specifications are written in such a way that it is impossible to carry them out.

Is Interpretation of a Contract a Question of Law?

A decision of the United States District Court for the Northern District of California may have an important bearing on future rulings of the Armed Services Board of Contract Appeals. As its name implies, this board deals with disputes between the government and contractors, including, of course, contractors for construction work. The court challenges the board's theory that its power to determine questions of fact precludes the courts from exercising jurisdiction, when in the court's judgment, the actual question involved is one of law.

In this particular case, the district court assumed jurisdiction even though the Board of Contract Appeals had ruled that the question involved was one of fact, thus making its ruling final. The court's opinion begins by the following review of the facts:

> Plaintiff is seeking damages for work he claimed he was ordered to perform pursuant to the terms of a contract containing no specifications for the said work. The contract dealt with the painting and reglazing of certain buildings located at McClellan Air Force Base, Calif.
> Plaintiff claims that although the contract made no provision for

the painting and reglazing of building designated by him as building
475F he was, nevertheless, required to repaint and reglaze building
475F and that the Government has refused to pay the contract price
for the work and the materials expended and placed by him in and upon
the building. He claims to be damaged in an amount equal to the contract price for the work and materials furnished to building 475F. The
problem presented to the court is solely one of interpretation of contract.

The contract included by reference the Government's Standard
Form 23-A, General Provisions, which provides for the settlement of
disputes through submission to the Contracting Officer and then by
appeal to the head of the agency involved, in this instance the Armed
Services Board of Contract Appeals. Both procedures were pursued by
the plaintiff, each resulting in a decision for the defendant United
States.

The Government has proceeded on the assumption that the determination made by the Board of Contract Appeals was one involving a
question of fact, and they rely on the provisions of the Wunderlich Act
(41 *U.S.C.* Secs. 321,322) and cases decided pursuant thereto. The Act
provides in substance that unless the findings of fact determined at the
administrative level are fraudulent, capricious or arbitrary, the court here
is bound thereby.

At this point the court took issue with the Board of Contract Appeals.
It said: "This court does not agree with the contention that the problem here
presented is one of fact. The interpretation of a contract is a question of law
(*Allied Paint and Color Works, Inc. vs. United States,* 309 F.2d 133 2d Cir.
1962) and regardless of any legal conclusions reached by the Administrative
Board below this court is free to make its own determination on all questions
of law (*Kayfield Construction Corp. vs. United States,* 278 F.2d 217 2d Cir.
1960)."

Having thus emphasized its disagreement with the Board of Contract
Appeals, it might be expected that the District Court would reverse the Board's
ruling and decide the case in favor of the plaintiff. But it did nothing of the
sort. After carefully examining the applicability of the contract, it granted
the government's motion for summary judgment, thus rejecting the plaintiff's
claim.

That claim, as indicated above, was that the plaintiff had been required
to repaint and reglaze building 475F although not required to do so by the
contract. The court therefore instituted a careful search through the specifications for building 475F and, as a result of this search, came to the conclusion that building 475F was a sort of figment of the plaintiff's imagination,
and actually was part of building 475 that both parties conceded the plaintiff
was required to repaint, reglaze, and do certain other work including replacement of space heaters, etc. The court recounted its search:

As previously stated, plaintiff alleges that he has performed work in excess of that required by the contract. The complaint asserts numerous times that plaintiff was required by the Government to perform work on building "475F" and that this building was not designated "as such" either in the contract or the specifications.

The court, in taking notice of the opinion rendered by the Armed Services Board of Contract Appeals, observes that the Board was unable to make a finding that plaintiff was required to perform work on building 475F. Instead, the Board found that what plaintiff referred to as building 475F was in fact building 475 and the contract was plain on its face in its requirement that plaintiff perform work on said building 475. In looking to the contract itself the court notices, as did the plaintiff and the Board, that it makes no reference to building 475F.

The contract is quite long and involved but all of the sections which are pertinent to the question now before the court were set out succinctly by the Armed Services Board of Contract Appeals in their opinion. The contract is quite free of ambiguity and the court will, therefore, mention only a few of the more relevant provisions.

Paragraph Ti-03, Project Requirements, sub-paragraph j, provides: "Project SMA 112-2;

"Building 475: Whip sandblast all concrete surfaces prior to re-painting; paint entire exterior surfaces including metal dust collectors, ducts and other metal work except cement asbestos siding; roof ventilators where shown on drawings are to be painted; replace in kind broken window panes; re-putty window panes as necessary; replace in kind damaged cement asbestos shingles. In building 475E, which is a portion of building 475, replace 19 space heaters, valves, strainers, traps, and fillings as shown on drawings."

Building 475 is well defined by Sheet No. 2 of the drawings, which is a part of Schedule "B" of the contract. Sheet No. 2 is entitled "Building Location Plan," building 475 being designated thereon.

The above is not ambiguous, but if any doubt remains, it is completely removed by reference to Sheet No. 4 of Schedule "E" of the contract. On Sheet No. 4 is found an irregular outline designated "Roof Building 475." A large section of the sketch is labeled "475." Other smaller and adjacent areas are labeled "475A" through "475E." On the same sheet is a list of what is to be done by the contractor.

At this point the court inserts a paragraph in tabular form showing that building 475 includes buildings 475A, 475B, 475C, 475D and 475E, with no mention of building 475F. The court continued:

Finally, in Section 1 of the specifications entitled Project Provisions, at paragraph P-03, subparagraph b, it is stated:

"The specifications and drawings which form a part of this contract are integral. Work shown on one, although not shown on the other, shall be executed as though shown on both. Should specifications and drawings conflict the specifications shall govern. Any doubt that may arise as to the intent and purpose of the specifications and drawings, shall be referred to the Contracting Officer."

The quoted specifications and drawings relating to the work in question are not in conflict nor are they ambiguous. The court is not required to look outside of the contract for an identification of building 475. It is clearly shown on the drawings which are part of the contract. Plaintiff's claim that he was not bound by any term in the contract to perform work upon a building designated "475F" is clearly correct. It is equally obvious that by the plain and clear words of the contract he was bound to perform the work he actually did perform on building 475, for which work he now seeks damages in excess of the contract price.

Having failed to find building 475F anywhere but in the plaintiff's imagination, the court granted a summary judgment in favor of the government. (*Crowder vs. United States*, 255 F. Supp. 873.)

What Happens When There Are No Plans or Specs?

Probably there is no better way to demonstrate the value of carefully prepared and easily understandable plans and specifications for a construction job than to relate what happens when owner and contractor try to get along without them. A case decided by the Superior Court of Pennsylvania was the fruit of just such an attempt. Toward the close of its opinion the court took occasion to remind the parties that a clearer understanding of what they were aiming at when they made the contract, plus detailed specifications, would have spared them a great deal of trouble.

The plaintiff in the action was a corporation that had agreed to install heating and air conditioning systems in a warehouse owned by the defendant. The contract was in the form of a letter of proposal from the plaintiff which the defendant-owner signed following the word *accepted*. The proposal used the phrase "we will furnish and install" followed by the trade name of the equipment which the parties intended to be installed. At the suggestion of the defendant-owner a guarantee was added, which read: "We guarantee to heat said building to a temperature of 70 F in zero weather." The price mentioned in the original proposal was $15,515, but this figure was crossed out and the figure $14,315 substituted. The letter began with the words: "As per your request we are pleased to requote on the above mentioned project."

The heating system was installed. Work had commenced in November, 1959, but differences had developed between the parties by the time the first test was made. The situation worsened over a period of several months and the air conditioning system never was installed by the plaintiff.

When the plaintiff decided to sue the defendant–owner, $7,000 had already been paid by the defendant on account. The plaintiff sued for $2,321.30 allegedly due for materials and labor, plus $3,091 for loss of profit. The defendant–owner filed a counterclaim for the sum of $3,797 based on the plaintiff's failure to perform the contract and the cost incurred by the defendant in correcting the heating system. The amount proved and submitted to the jury was $2,242.35.

The jury at the trial in the Court of Common Pleas, Lehigh County, brought in a verdict which amounted to a standoff. The court thereupon entered judgment for the defendant on the plaintiff's claim and for the plaintiff on the defendant's counterclaim. The plaintiff appealed to the Supreme Court.

That court not only affirmed the judgment of the trial court but adopted the opinion of Judge Henry B. Scheirer of the lower court as its own. Excerpts from Judge Scheirer's opinion show the difficulty of the task presented to both judge and jury as the result of the failure of the parties to prepare and use plans and specifications which would have made their intentions clear to each other. The dispute that led up to the lawsuit is described in Judge Scheirer's opinion as follows:

> Work was begun by the plaintiff on November 13, 1959 and, according to the defendant, substantially completed by December 13, 1959. The plaintiff does not agree that the installation was substantially completed on December 13 and we shall present the reasons later. The warehouse was almost completed either by December 10, 1959 or December 18, 1959. The letter of proposal was brief and lacking in detailed specifications. Nor was there an exact time mentioned when the guarantee to heat the building at 70 F in zero weather was to take effect. Considerable conflict occurred over the location of the unit heaters. Counsel for plaintiff in his brief said, "It is important to note at this point that C. Lewis Stryer is a builder with vast experience, and it is inconceivable that Mr. Stryer should sign a contract for the installation of a $14,000 heating and air conditioning system without approving plans showing the *design of installation*." It seems clear, however, that the defendant did sign the contract without approving a "design of installation."
>
> The testimony of the plaintiff is that the defendant submitted a plan of the building which had the location of the unit heaters marked for the center of the warehouse. The plan was prepared by an engineer,

not called as a witness, but who informed the court following our inquiry that it was to be used for the purpose of securing approval of the Department of Labor and Industry and that the heat unit locations were not placed on the plan by him. Further, the height of the heaters was not indicated on the plan. There was no reference in the contract to the plan. The defendant made a statement that is contradictory to the plaintiff's position that he, the defendant, knew the "design of installation." He said,

"A. Well, Allentown Supply never gave me any indication of where they were going to hang these heaters. I took it for granted they knew what they were doing and since they made a guarantee of 70 F at zero, I wasn't too much worried as to what their activities were going to be. So I happened to come over one morning and I saw them hanging the heaters down about 12 ft from the floor."

"Q. About when was that please?"

"A. Approximately November 13."

"Q. 1959?"

"A. 1959."

After three unit heaters were installed in the center of the building at a height of 13 ft [it was testified by plaintiff that hanging the units in this manner was common practice], the defendant said he requested that the heaters be placed on the perimeter of the building because of his view that the suspended heaters with steam pipes below them would limit the use of the floor to one side only. The defendant testified that the distance between the steam pipes and the floor would be about 8 or 9 ft. The defendant further expressed his doubts as to whether the building could be heated to a temperature of 70 F as guaranteed. Mr. Wasser, an employee of the plaintiff, testified that the defendant requested the units to be raised. Thereafter, the plaintiff raised the heaters to a height of 16 to 17 ft prior to December 10. As stated before, the heating system was substantially completed by December 13.

The air conditioning work had not been instituted by this date. The floor of the warehouse had not been concreted by then and a tarpaulin hung over a door. On December 16, an engineer [Bowman] visited the premises at the request of the defendant. At the thermostat level (54 in.), he found the temperature to be 44 F. It was considerably less at floor level and "unbearable" in the ceiling area. It was this engineer's opinion that the building could not be heated to a temperature of 70 F with the units in the center. His recommendation was that the units be placed along the perimeter and cited the suggestion of a heating manufacturer [Kritzer] other than the one involved in the contract but whose heating units were installed. To do this would eliminate stratification and produce better circulation.

There was a clear difference of testimony between plaintiff's and defendant's witnesses as to whether the thermostat was ever set at 70 F. The plaintiff's witnesses testified they never saw the thermostat set at 70 F and one of the defendant's employees was quoted as saying that he turned the thermostat off at night to save fuel. The defendant himself testified that the thermostat was set at 70 F and the plant was run at that temperature continuously. The temperature then at 18 in. off the floor was 38 F. When the thermostats were placed at 4½ ft the temperature was 56 F to 58 F. At the ceiling level the temperature was much higher, possibly 100 F. The defendant further testified that at no time after December 10 [until the system was altered] did the temperature rise to 70 F. The outside temperature during the winter of 1959–1960 was termed "mild."

Relations between the parties went from bad to worse. The defendant called in an engineer who recommended an enlargement of the heating system with additional heating units both in the center and at the perimeter of the warehouse. The plaintiff called in another engineer who was denied admission to the warehouse, but who looked at the plans and at the trial testified that it was common practice to place heating units in the center of buildings used for warehouse purposes. Hostilities continued for several months thereafter, employees of the plaintiff testifying that they were barred from the building and the defendant testifying that he made repeated efforts to induce the plaintiff to correct the work. Finally, after legal counsel had been called in by both sides, the defendant terminated the contract on the ground that the plaintiff had failed to meet its guarantee. (*Allentown Supply Corp. vs. Stryer,* 195 A, 2d 274.)

Can Specs Be Impossible to Carry Out?

The Supreme Court of Colorado considered a case that, fortunately, is rare in the construction industry. The question: Was a set of job specifications impossible to carry out?

First, the court had to adopt a workable definition of *impossible;* second, it had to determine whether the defendant contractor had proved that the job could not be done according to the specifications. Many engineers were called to testify during the trial.

The contract was for building two 5-million-gallon water tanks for the City of Littleton. When the contractor, Latimer and Gaunt, defaulted, contending that the performance of the specifications was impossible, the city brought an action for breach of contract. The trial court rendered judgment for the contractor, and the city appealed to the state supreme court.

The facts of the case, according to the supreme court, were these:

On May 8, 1961, Littleton hired the contractor to build two precast and prestressed concrete water tanks. The city's plans for each tank called for a poured concrete base with a 175-ft dia. Around the edge, 92 precast wall panels, each 6 feet wide and 29 feet high, were to be set in a slot in the base and joined. Once in place, the wall panels were to be prestressed by wrapping wire cable under tension around the circumference of each tank. During construction, the tanks collapsed.

"Thereafter," the court said, "Littleton and the contractor entered into a Supplemental Agreement which, among other things, required reconstruction of the tanks pursuant to plans and specifications already in existence, and in accordance with instructions to be supplied by Littleton's consulting engineer. No work was done pursuant to the Supplemental Agreement. In an exchange of letters after its execution, the contractor requested but was refused additional written details regarding reconstruction."

On April 15, 1962, the contractor ended the agreement by telegram and the next day sent a letter to the city councilmen explaining that he had ended the agreement because the design was faulty. Littleton then brought suit, charging that the contractor had breached his obligations under the Supplemental Agreement. The trial court found that performance was impossible as a practical matter. Citing the Restatement of the Law of Contracts Sec. 454, the trial court ruled as a matter of law that the impossibility constituted a complete defense to the contractor's claims. Littleton contended that the contractor either knew or should have known of the facts that supported his defense of impossibility when it signed the agreement; that the city and contractor, by that agreement, manifested an intention that impossibility should not be a defense; and that, therefore, the defense of impossibility was inapplicable.

After thus stating the facts, the supreme court first took up whether the work specified in the contract was impossible. It began with this legal definition of impossibility:

> The law recognizes impossibility of performance as a defense to an action for breach of contract. In his brief, the contractor suggests that we adopt the Restatement Sec. 454 definition of impossibility. According to Restatement, impossibility means not only a strict impossibility but impracticability because of extreme and unreasonable difficulty, expense, injury or loss involved. . . . "Impossible" must be given a practical rather than a scientifically exact meaning. Impracticability rather than absolute impossibility is enough. Restatement Sec. 454 and comment thereunder.

Commenting on the Restatement Sec. 454, Williston has stated that:

"[t]he true distinction is not between difficulty and impossibility. A man may contract to do that which is impossible. . . . The important question is whether any unanticipated circumstance has made performance of the promise vitally different from what should reasonably have been within the contemplation of both parties when they entered into the contract. If so, the risk should not fairly be thrown upon the promisor." 6 Williston, Contracts Sec. 1931 (Rev. ed.) See 6 Corbin, Contracts Sec. 1325.

Other jurisdictions have begun to adopt the Restatement definition. See A.L.R. 2d12, and cases therein cited. After reviewing the authorities which have been subscribed to the Restatement definition of impossibility, we hereby adopt the definition of impossibility contained in the Restatement 454 as interpreted by Williston. See commentary on this section contained in Storke, Colorado Annotations to Restatement of the Law of Contracts.

Having defined impossibility, the court turned to whether the Littleton specifications met that definition. It said:

The trial judge in this case heard 15 days of testimony in a trial which commenced on December 14, 1964, and terminated on January 14, 1965. . . . The record reflects that the trial judge observed at least one demonstration in the courtroom. Throughout most of the trial, he listened to the expert testimony of consulting engineers. Thirty days after the trial ended, the court entered carefully considered and detailed findings of fact, conclusions of law and judgment. Under these circumstances, it is particularly appropriate to reiterate the general rule that the judgment of the trial court is presumed to be correct, and that the findings of the trial court are conclusive if supported by the evidence.

One of the engineers whom the trial judge heard said the construction of the tanks, as originally specified or as modified by the Supplemental Agreement, was not possible either as a structure or as a water–containing tank later. Another testified that he certainly would not try to reconstruct the tanks on the information supplied in the documents. Another contractor told Latimer that he thought it was impossible to construct the walls of the tanks as specified. According to Latimer's testimony, even the consulting engineer for Littleton admitted that he was "scared of these tanks." On cross-examination, Littleton's engineer admitted that the specifications on one tank would have to be changed to facilitate reconstruction. Although one engineer was apparently convinced that construction was physically possible, he appeared to be equally convinced that the tanks would collapse soon after construction.

"Without detailing the other testimony," the supreme court said, "suffice it to say that the consensus was that the tanks could not be reconstructed as contemplated under the Supplemental Agreement consistent with sound engineering practice. We conclude that there was substantial evidence in the record to support the trial court's finding that the tanks contemplated by the Supplemental Agreement were impossible to build either physically, as structures, or functionally, as water–containing tanks, within the Restatement definition of impossibility of performance."

The supreme court then turned to the city's contention that the contractor made the agreement with such knowledge of the situation that it foreclosed the defense of impossibility of performance. On this question the court said:

> According to Littleton's argument, when the contractor entered into the Supplemental Agreement, it knew or had reason to know of the facts which later furnished the basis for the trial court's finding of impossibilility of performance. If the evidence supports this assertion, the impossibility of performance would be no defense to Littleton's case.
>
> The Supplemental Agreement, which had been submitted to the contractor by Littleton, was signed by the contractor on November 11, 1961, and by Littleton on Dec. 5, 1961. Since Littleton's argument bears directly on the knowledge which the contractor had when it executed the agreement, we will treat November 11, 1961 as the significant date.

The court noted that, aside from expressly providing that construction should proceed in accordance with instructions from the city's consulting engineer, the agreement outlined the scope of the work "subject to modifications as the work progresses." Moreover, the court said, on direct examination, Latimer testified that when the agreement was signed, he "had a great deal of faith" in what the consulting engineer had to say about reconstructing the tanks. The record clearly supports the observation that after the tanks collapsed, the city and the contractor knew that some modifications of the original specifications would become necessary, the court said. But that evidence, the court added, "also supports the further inference that on Nov. 11, 1961, both parties to the contract expected that construction would proceed. We cannot, therefore, say that as a matter of law the trial court was required to find that the contractor undertook what it knew was impossible to accomplish when it signed the Supplemental Agreement."

The trial court also found that performance was impossible due to insufficient and defective plans, specifications, and directions. As one of the briefs put it, even the engineers had their consulting engineers. Extensive testing was begun to discover the cause of collapse.

Noting all that, the supreme court said:

> The record is complete with evidence to support the proposition that conclusions as to the cause of the collapse were not reached until well after the Supplemental Agreement was signed. For example, Latimer testified that after signing the agreement he talked to other engineers and then came to the conclusion that the tanks could not be built according to any plans which had been furnished to that date.
>
> The consulting engineer for Littleton theorized that the basic cause for the collapse was an absence of grout in the joints and misalignment of the panels. But he testified that he had not reached a conclusion as to the cause of the collapse until March of 1962, over a year after the Supplemental Agreement was signed.
>
> Twice in April of 1962 as the trial court found, the contractor requested additional instructions, and was told that the instructions given were sufficient. Only after Littleton once more demanded that reconstruction begin, without supplying additional instructions, did the contractor terminate the agreement.

In passing, the court noted that after Latimer and Gaunt ended their agreement, Littleton redesigned the joint between the panels, modified the original specifications some more and relet the project for public bid. But no bid was received on the precast method of construction. Steel tanks were eventually erected on the concrete slabs originally poured to provide the bases for the concrete tanks.

The court then discussed Littleton's suggestion that the parties expressed an intention that impossibility should be no defense to an action for breach of contract. The Supplemental Agreement provided, first, that the contractor was not responsible for the engineering or design of the tanks and, second, that Latimer and the city wanted to avoid litigation to determine the responsibility for the collapse of the tanks. According to Littleton, those contract terms showed the intent of the parties to be that impossibility of performance should be no defense. So Littleton argued that the contractor should have tried to erect water tanks that, according to the testimony of experts, either could not be erected or, once erected, could not stand. Having failed to try, Littleton contended, the contractor could not say his obligation was discharged by impossibility.

"As Littleton argues," the court continued, "the defense of impossibility is inapplicable where the contrary intention has been manifested. Restatement Secs. 456, 457. The exception to the defense of impossibility is inapplicable where, on an interpretation of the contract in the light of accompanying circumstances and usages, the risk of impossibility due to presently unknown facts is clearly assumed by the contractor. In view of the contractor's express disclaimer of responsibility for defects in the design in the Supplemental Agreement, there can be no serious contention here that the . . . contractor intended to assume the risk. Restatement Sec. 456, Illustration 4. Nor does the recital that the parties desired to avoid litigation raise an inference that the contractor assumed the risk of impossibility."

The court continued:

> There is no merit in Littleton's argument that defective specifications cannot provide the basis for the contractor's claim that performance was impossible. When the Supplemental Agreement was executed, both parties contemplated the necessity for additional instructions from the city's consulting engineer. In this case, Littleton's insistence that the instructions were adequate, and its corresponding refusal to issue the additional instructions requested by the contractor, later provided the basis for the defense of impossibility of performance. Without additional instructions, as the engineers testified, performance was impossible.
>
> We have not overlooked Littleton's argument that the Supplemental Agreement was a compromise agreement and that therefore the contractor was precluded from setting up defenses which might have been available to him if Littleton had initiated suit immediately after the tanks collapsed, instead of entering into the Supplemental Agreement. We are not dealing here with defenses possibly available to an earlier agreement which have been compromised by a subsequent agreement. Instead, we are concerned with a new contract whose terms require reconstruction of the tanks pursuant to instructions from the city's consulting engineer. The circumstances of the execution of the Supplemental Agreement do not suggest that the contractor intended to relinquish the right to raise the defense of impossibility to an action brought against him for an alleged breach of the Supplemental Agreement. Further, the Supplemental Agreement required that instructions from the consulting engineer be in accord with good construction and engineering practices. As the expert witnesses testified, however, it was impossible, within the meaning of the word as we have defined it, for the contractor to reconstruct the tanks in accordance with the instructions furnished him. Thus impossibility of performance was a valid defense in the present action.

The judgment of the trial court in favor of the defendant contractor was affirmed. (*City of Littleton vs. Employers Fire Insurance Co.*, 653 P.2d 810.)

Were the Specs Faulty—or Was the Workmanship Poor?

The United States District Court in Rhode Island decided a lawsuit in which the plaintiff, a contractor, contended that the failure of his work was due entirely to defective specifications, in particular, a requirement that a certain material be used. The defendants attributed the failure to defective workmanship on the part of the contractor. The court, in awarding more than $100,000 to the contractor, put the responsibility on the defective specification, which prescribed a mortar material that, as it turned out, could not do the job.

The claim was for payment due under a contract for building a system of underground piping. The defendant, in a counterclaim, sought liquidated and other damages. Fanning & Doorley Construction Co., the plaintiff, is a Rhode Island corporation. The defendant, Geigy Chemical Corp., is a Delaware corporation with main offices in Ardsley, N. Y. Geigy has a plant in Cranston, R. I., the Alrose Division, which manufactures chemicals.

In 1958, Geigy decided to build a new system of underground piping at its Alrose plant. It hired Metcalf & Eddy, consulting engineers from Boston, to draw plans and specifications, prepare a contract, let the contract, and supervise its execution. Accordingly, the engineering firm prepared the "book," which contained, among other things, information for bidders, form of bid, contractor's bond, and specifications.

The work under the proposed contract consisted of cast iron water piping, concrete storm drains, a vitrified clay sanitary sewer system, an industrial waste sewer system, and a booster pumping station. Metcalf & Eddy, after an investigation by one of its engineers, John Podger, recommended (and it was so specified) that the industrial waste system be made of chemical stoneware bell-and-spigot pipe, and that the pipe be joined by a material called Causplit, a mortar produced by blending a resin and powder that is resistant to the corrosive effects of the acids, alkalis and alcohols that the Alrose plant discharges.

In July, 1959, the book was circulated and bids were invited. In the section titled "Information for Bidders," prospective bidders were warned to satisfy themselves on the conditions and requirements of the work. It also said that the information furnished was guaranteed to be accurate. In addition, bidders were told to satisfy themselves regarding the character, quantities, and conditions of the various materials to be used.

Fanning & Doorley, on July 31, 1959, submitted its bid on a unit-price basis using the form set forth in the "book." In this form of bid, units and quantities of different kinds were estimated and a price submitted for each. Fanning & Doorley's bid on the units estimated in the book totaled $204,147.50.

The prescribed contract was executed on Aug. 28, 1959, by the appropriate officers of Fanning & Doorley and Geigy. At the same time, a Supplemental Agreement was signed; it provided, among other things, that Fanning & Doorley be paid on the basis of its cost plus 15 percent but that the total amount payable would not exceed the amount computed by using the prices stipulated in the proposal. In other words, this was a cost-plus contract with a top, or upset, limit.

After the contract was signed, the work was scheduled to begin on Sept. 21, 1959, and be completed by Feb. 21, 1960, the end of the 180-day specified contract period.

The pertinent specifications for the chemical stoneware pipe were as follows:

> Except as hereinafter specified, the pipe shall be joined with asbestos-rope caulking and Causplit mortar made by the Pennsalt Chemical Corp., Philadelphia. The contractor's attention is directed to the fact that Causplit is a phenol–formaldehyde type resin and that some persons are allergic to such materials. Causplit should at all times be handled, and all joints shall be made, with the recommendations of the manufacturer, and extreme care shall be taken in its use to avoid any possible damage.

> Expansion joints shall be made, as indicated on the drawings, with asbestos–rope caulking and a plasticized epoxy similar to Caroline E manufactured by the Ceilcote Co., Cleveland.

> After the pipe has been laid and jointed but before the concrete cradle is placed, all areas of the pipe that would otherwise be in contact with the concrete shall be covered with a bituminous impregnated felt to ensure that a bond will exist between the pipe and the cradle.

In addition, there was a provision that the sewers and manholes had to meet certain leakage test requirements, and that, "Portions of sewers which fail to meet tests shall be repaired, and retested as necessary without additional compensation until test requirements are met."

The agreement further stipulated that the engineering firm would in all cases determine the amount, quality, acceptability, and fitness of the work, and that if the contractor questioned a decision of the engineers, the decision of the engineers would be a condition precedent to the contractor's right to receive money for the work to which the question or difference in opinion related.

Article II of the agreement provided that the contractor would do everything that the contract required in the manner and within the time specified and that the entire work would be completed to the satisfaction of the engineer.

Article XXIV provided that, "If the work or any part thereof shall be found defective at any time before the final acceptance of the whole work, the contractor shall forthwith make good such defect in a manner satisfactory to the engineer. . . .”

Fanning & Doorley started work under the contract in the latter part of September, 1959. Afterward, Metcalf & Eddy, on the authority of Geigy, issued 12 written supplemental agreements that provided in some cases for extra work or a redesign of work and in other cases a deletion of work.

The defendant contended that the major item not completed by July, 1960, was the chemical stoneware system. The defendant said this item was crucial, and it was apparent that Fanning & Doorley had not completed and was not going to complete this requirement.

Fanning & Doorley began laying the chemical stoneware pipe and making Causplit joints about Dec. 1, 1959. Although it had no experience in this kind of work and did not make contact with the manufacturer or, at the outset, make a test joint, it did seek guidance from Geigy and was told that the work would be under the direction of the resident engineer of Metcalf & Eddy.

When the first joint was made, it was done under the direction of resident engineer John Elwood, who mixed the material and either helped or troweled it into the joint. Elwood testified that thereafter the recommendations and orders or suggested changes of the manufacturer were carried out at all times by Fanning & Doorley; that the joints were made under Elwood's supervision; and that weekly reports were sent to Podger, the engineer who did the original investigating for Metcalf & Eddy. Significantly, Elwood also stated that the trenches in which the pipe was laid were backfilled at either his or his superior's direction. All joints made up to March 23, 1960, failed to pass leakage tests.

Here the court said:

> The daily history of the use of Causplit mortar as testified to by John Elwood and as reflected in the exhibits logging each day's activities develops a series of trial and error showing a lack of the required specific knowledge as to its proper use, procedures and application on the part of the defendant's engineers and the manufacturer of Causplit.
>
> It would be inappropriate for this court to detail over 3,000 pages of testimony in support of each of its findings. However, certain significant points should be noted. Starting in December of 1959, Fanning & Doorley followed the directions and recommendations made to them as to the application of Causplit mortar. All that they did was supervised and documented by the defendant's resident engineer who, as leakages occurred in the various joints and difficulties were encountered, directed new and different procedures.

Part of this recorded history, the court said, showed the following changes in procedures, all of which Fanning & Doorley were told to do:

On Dec. 1, 1959—simple troweling was employed; Dec. 18—because of water condition, the drying of pipes and application of heat were directed by the defendant; Dec. 21—cradling pipe with concrete, coating it with asphalt and piping heat were directed by the defendant, and this was further modified on Dec. 22; Dec. 31—the manufacturer replaced the Causplit mortar with one having a faster setting ingredient and recommended the use of space heaters.

Jan. 13, 1960—a representative of the manufacturer suggested that more powder be added to the mix to increase the setting time and sulfur cement be poured around the Causplit (it was understood that this would eliminate the need of pumping out the wet trenches and the application of heat to the pipes to cure the mortar); Jan. 25—Fanning & Doorley received the reprocessed mortar and began applying it.

Feb. 1, 1960—to overcome the difficulty of keeping the mortar in the joints, the manufacturer suggested placing dry ice under the mortar pan; Feb. 2—a representative of Pennsalt, the manufacturer, again called at the job site, directed the making of a sample joint and, in doing so, found that the mortar wasn't workable—it slumped out of the joint—whereupon the defendant again changed the procedures, this time in the caulking, following which the mortar seemed to work; Feb. 26—a representative of Pennsalt again went to the job site and tested the joint made under his direction and found the leakage exceeded specification requirements.

In March, 1960, the joint the manufacturer made had an even greater percentage of leakage. On March 10, Metcalf & Eddy, through Podger, the project engineer, told Fanning & Doorley to stop laying the chemical stoneware sewer and had Douglas Woolley, a factory representative of Pennsalt, come to the job site. Employing new techniques, he finally devised a procedure for making leak-free joints. Part of this new technique was the use of different caulking material.

It was at this point that a successful method of making leak-proof joints was first devised and demonstrated to Fanning & Doorley. But even this joint, made under Woolley's supervision, had void spaces in the top part, and as Elwood testified, this indicated the leakage was being prevented by the asbestos rope, which had to be bought from other specified manufacturers than Pennsalt—that is, a new material for caulking not previously specified.

The court continued:

> In connection with this testimony, the court must look to the contract for the original specifications set forth therein and the contract obligation of the plaintiff.
>
> Section 16.4 is written in positive terms with absolute requirement that asbestos rope caulking and Causplit mortar made by Pennsalt be used in making chemical stoneware pipe joints. This is different from the specification requirements for every other type of joint where a

degree of discretion is permitted. It must be noted that this section also has the mandatory provision that the joints ". . . *shall* be made in accordance with the recommendations of the manufacturer. . ." [Emphasis added].

It can hardly be said that the plaintiff should be held responsible for leaking joints caused by the use of Causplit mortar as recommended by the manufacturer.

The defendant argues that the plaintiff's work was defective in that, among other things, certain pipes were cracked, void spaces were found in the mortar of many joints, improper caulking and fitting of the pipes. This is not acceptable as proof of poor workmanship or as a contributing cause of the leakage. The testimony shows that the cracks may well have been caused by a backhoe which was equipment used by the defendant in excavating the ground to expose the pipes for inspection or improper grades as set by the defendant's engineer and as to the void spaces, the manufacturer could not prevent this as evidenced by every joint it made.

The defendant introduced oral testimony of a caulking iron being found in a joint of a 12-in. line. It seems strange that this evidence, documented by photographs to show poor workmanship, did not reveal to the satisfaction of this court any such tool. In short, evidence of poor workmanship in the laying of pipes up to March is not convincing. [The court finds] it difficult to attribute the quality of the work as to void spaces and lack of caulking to Fanning & Doorley—these very faults were proven to be attributable to the inherent failure of the materials and procedures imposed on the plaintiff. Nor can this court attach any significance to the defendant's assertion that the plaintiff prematurely backfilled the trenches after the pipes had been jointed and permitted their use for disposal of chemical waste. The short answer to this is the order to do so was solely in the defendant's engineer and the owner to the exclusion of the contractor. This court finds that: A. All joints made up to March when Mr. Woolley finally devised the procedure for making a leak-proof joint, were made by the contractor following the specifications and in strict accordance with the directions, orders and recommendations of the engineer and the manufacturer; B. that up to March, 1960, it was not possible to make a tight joint following the procedures and directions theretofore specified; C. that the specifications required the contractor to do work in a manner and to a degree which could not produce the desired result as originally written insofar as the chemical stoneware system for the disposal of industrial waste was concerned. The specification requirements and all specific directions, recommendations and orders prior to March 23, 1960, were defective, inade-inadequate and faulty.

Having ruled that Geigy was responsible for the failure of the work, the court decided that the amount due the contractor was $103,414.31. (*Fanning & Doorley Construction Co. vs. Geigy Chemical Corp.*, 305 F. Supp. 650.)

How Can Plans and Specs Cause Construction Delays?

The dire consequences of defective plans and specifications are strikingly illustrated in an exhaustive opinion handed down by the United States Court of Claims.

The contractor on a job calling for the construction of certain aircraft maintenance facilities at Willow Grove, Pa. (the original contract price was $1,480,112 which was later increased by change orders to $1,700,166.50) brought suit against the United States for the sum of $248,665.76 as compensation for certain additional items of the work which it had performed.

The chief cause of the claim was the fact that the completion of the job took place 518 calendar days beyond the scheduled completion date. Although the government extended the completion date by 518 days and did not exact the penalties provided for in the contract, the contractor nevertheless contended that the defective specifications were the principal cause of the delay and the additional work which it had had to perform. Included in the list of claims were claims for additional home office overhead and loss of productivity directly due to the delay in the completion of the job.

The action was brought after various government agencies, the contracting officer, the general accounting office, the Secretary of the Navy, and the Armed Service Board of Contract Appeals, acting for the Secretary of the Navy, had all rejected the plaintiff's claims.

In its opinion the Court of Claims stated the plaintiff's claims as follows:

On November 9, 1959, the plaintiff filed a petition in this court asserting that the defendant breached the contract as follows:

(a) The original plans were defective and faulty; (b) defendant was dilatory in making necessary contract changes and taking other action; (c) a trial and error method was imposed upon plaintiff by defendant to accomplish certain aspects of the construction and this was contrary to the contract terms and was unreasonable and costly to plaintiff; (d) the magnitude and nature of some of the changes were beyond the scope of the original contract; and (e) the frequency and extensiveness of the revisions to other phases of the construction work were also unreasonable and beyond the scope of the contract. As a result of these breaches, plaintiff contended that its costs of construction were increased and performance of the contract was protracted unduly and into periods of more adverse conditions than would otherwise have occurred.

After an extensive trial, the court's trial commissioner ruled that the defendant unreasonably delayed the plaintiff in the performance of its contract a total of 420 days out of the total 518 days of the overrun period. The commissioner also determined that as a result of these unreasonable delays the plaintiff had incurred delay costs for idle equipment, field supervision, winter protection, rehandling materials, maintaining excavations, wage and material price increases, and additional insurance premiums.

The parties accepted the trial commissioner's report with the following exceptions: The defendant government contended that only 300 days instead of 420 should be considered as unreasonable delays attributed to the defendant. The plaintiff contended that the trial commissioner erred in not allowing two items of damages—excess of home office overhead and loss of productivity of its labor force. The court found that the defendant's contention was without merit but disagreed with the trial commissioner's refusal to allow damages for excess of overhead and loss of productivity.

The court's opinion proceeded to a detailed consideration of the faulty plans and specifications and the additional work created by their inadequacy. That portion of the opinion occupies about three pages and will not be quoted here as the exact nature of the defects does not affect the legal principles which determined the outcome. It is sufficient that they were deficient and greatly increased the work which the plaintiff had to perform.

At the conclusion of this lengthy recital of the facts the court said:

> That the original specifications were defective is beyond dispute. They misrepresented the nature of the bearing value of the material underlying the foundation of the structure at the prescribed elevations and hence the dimension and depth at which the arch-column footings were to rest to such an extent that the changes required to complete the structure were beyond the scope of the original contract.
>
> It is well settled that when the Government orders a structure to be built, and in so doing prepares the specifications prescribing the character, dimension and location of the construction work, it impliedly warrants that if the specifications are complied with, satisfactory performance will result. [Citations] When, as here, defective specifications delay completion of the contract, the contractor is entitled to recover damages for defendant's breach of this implied warranty. [Citations]
>
> In addition, the defendant was dilatory both in recognizing the need for and in making appropriate revisions to the defective foundation plans. Defendant should have determined whether the subgrade rock had adequate bearing capacity prior to approving the pouring of the first D-line footings. Upon finally recognizing that the subgrade rock was unsatisfactory, defendant or its agents should have completed the redesign of the foundation with all due haste so that plaintiff could have continued the foundation work without any significant delay.

Furthermore, the defendant should not have imposed upon the plaintiff the unreasonable trial and error method of excavation once the revised design had been completed. This requirement, together with the extremely slow recognition and correction of the defective plans, constituted a breach of the implied obligation contained in every contract that neither party will do anything that will hinder or delay the other party in performance of the contract. [Citations]

That these breaches of contract by defendant caused considerable delay in connection with the foundation work is clear from the foregoing. Plaintiff is of course entitled to receive the damages it incurred because of this delay.

Upon consideration of the evidence our trial commissioner has determined that the defendant unreasonably delayed the progress of the foundation work for a period of 330 days of the total 420 days' delay. This finding is fully supported by the record.

After considering the impact of some strikes which occurred during the progress of the work, the court turned to the contention that the defendant took too much time to make necessary changes once the defects had been discovered:

Defendant contends that the trial commissioner's allocation of 330 days to unreasonable foundation work delay should be further reduced by 30 days which, according to defendant, should have been allowed as a reasonable length of time to make the necessary design changes once it was determined that the original elevations for the arch–column footings were not acceptable.

Ordinarily defendant is entitled to make necessary changes, but where the change is necessitated by defective plans and specifications defendant must pay the entire resulting damage without any reduction for time to make changes, as would be the case if the redesign was necessitated by a changed condition or the like. [Citation]

In summary we reach the conclusion that the defendant unreasonably delayed the plaintiff in the performance of the contract work by a total of 420 days out of the 518-day overrun period.

The trial commissioner has determined that the plaintiff suffered delay damages and additional expenditures directly attributable to the defendant's breaches in the amount of $85,544.92. This amount is made up of $\frac{430}{518}$ of the overrun period costs for idle equipment, field supervision, winter protection, maintaining excavations, and wage and material price increases plus 100% of an insurance premium plaintiff was required to pay. This fairly represents the damages suffered by plaintiff with respect to these items.

The court then took up the disputed claims for home office overhead and loss of productivity of its labor force. In regard to the first of these it said, "We are of the opinion that the trial commissioner was in error in finding that plaintiff was not entitled to home office overhead and hence in failing to determine the amount thereof. Home office overhead is a well-recognized item of damage for delay and plaintiff would be entitled to recover it, if it did not release its claim to it. [Citations]"

The court overruled defendant's contention that plaintiff had released this claim. However, it had considerable difficulty in arriving at the amount that plaintiff should receive for home office overhead.

The court also had difficulty in arriving at the amount that the plaintiff should receive for loss of productivity of its working force although it disagreed with the trial commissioner who had denied recovery for this item. The court fixed the amount at $16,589.42.

The opinion concludes with a paragraph awarding damages to the plaintiff, "We have found that the plaintiff is entitled to recover $62,948.33 for excess home office overhead. Adding these two items to the $85,544.92 found by the trial commissioner makes a total of $165,082.67. A judgment for this amount is entered in favor of plaintiff against defendant." (*Luria Brothers & Co. vs. United States,* 369 F.2d 701.)

Does Everything Have to Be in Writing?

The wisdom of having the written contract provide for every foreseeable contingency that may arise on a construction job is shown in a case decided by the Supreme Court of Mississippi.

A contractor brought suit against a bank for alleged extra work not included in his contract, contending that the work had been ordered by the structural engineer employed by the architect. The trial court awarded the contractor $8,902.21 on a *quantum meruit* basis for the extra work, but the bank appealed to the Supreme Court of Mississippi. That court reversed the ruling and held that the alleged extra work was included in the written contract for a fixed sum.

(The Latin legal phrase *quantum meruit* crops up frequently in this case, and so a few words explaining its meaning may be in order. Where an actual contract for work performed cannot be proved, the law does not permit the party who has received the benefit of the work to get off without paying for it. It invokes the *quantum meruit* theory, which means that the party who is benefited must pay to the party who has performed the work the amount the court decides, even though no agreement for such payment was entered into by the parties.)

Here are the details of the case:

In 1964, the bank and contractor entered into a written contract for demolishing the existing bank building and for building a new bank on the same site in Meridian. After the old building was removed, the contractor, testing the site for placing concrete piling for the foundation of the new building, encountered the foundations of the old building. The foundations were some five feet below the surface of the basement of the old building. Although their presence had been anticipated, their location and size had been unknown. Their location obstructed the placing of the new piling, necessitating that they be either bored through or removed.

The contractor asked the structural engineer, an employee of the architect, whether some of the old foundations could be used for the new structure. Told that this was infeasible, the contractor removed the foundations, which entailed much labor and expense but which was, nevertheless, cheaper than boring through them. (A conflict in testimony arose between the engineer and the contractor. The engineer testified that he told the contractor that it would be necessary for him to remove only that portion of the existing foundation as would make room for the new piling. The contractor's witnesses testified that the engineer directed that all the old foundations be removed.)

While the foundations were being removed, the contractor advised the bank's architect that he considered the work supplementary and that he expected more payment. Accordingly, when the work was completed in May, 1965, he presented a bill to the architect for this work. The architect testified that he never directed the removal of the old foundations and that he was unaware the contractor expected additional pay until shortly before the bill was presented. They discussed the matter on several occasions, and the architect apparently thought at first that the contractor was entitled to be paid more than the contract price for the work performed. He told the contractor that he would help him obtain this payment. Upon re-examining the contract, however, he concluded that the work was not extra and was within the contract price. He therefore declined to certify to the owner that the contractor should receive reimbursement for extra work.

After a long court hearing in which all contract documents were introduced, the hearing judge found the bank liable on a *quantum meruit* basis for the removal of the old foundations. He found "that the owner [the bank] personally through its board of directors never officially passed on this subject and certainly never signed anything or gave any oral statement to the effect that they agreed this was an extra. There is evidence that the architect agreed that this was an extra, but changed his mind, and there is no doubt . . . but that from the beginning of the work the contractor felt that this was an extra."

The hearing judge also found that the owner had not waived any provisions of the contract.

The supreme court identified the relevant portion of the contract as:

2-F—CONCRETE PILING

(a) Scope: *Furnish labor and materials to complete bored, cast in place concrete pilings* as indicated, specified herein or both. . . .

(4) Equipment: Use drilling equipment generally used in standard pile boring practice as approved.

(5) If obstructions such as masonry (sic) *old foundations,* etc., are encountered, bore them as directed.

(6) Payment:

a. Except as herein provided, *no separate payment will be made for specified work; include all costs in connection therewith in Stipulated Sum for entire work under contract.* [Emphasis added]

Attached to the contract and made a part of it was a pamphlet prepared by the American Institute of Architects titled "General Conditions." Article 15 is as follows:

CHANGES IN THE WORK

The owner, without invalidating the Contract, may order extra work or make changes by altering, adding to or deducting from the work, the Contract Sum being adjusted accordingly. All such work shall be executed under the conditions of the original Contract except that any claim for extension of time caused thereby shall be adjusted at the time of ordering such change.

In giving instructions, *the Architect shall have authority to make minor changes in the work, not involving extra cost,* and not inconsistent with the purposes of the building, but otherwise, except in an emergency endangering life or property, *no extra work or change shall be made unless in pursuance of a written order from the Owner,* signed or countersigned by the Architect, or a written order from the Architect stating that the Owner has authorized the extra work or change, *and no claim for an addition to the Contract Sum shall be valid unless so ordered.* . . .

Should conditions encountered below the surface of the ground be at variance with the conditions indicated by the Drawings and Specifications the Contract Sum shall be equitably adjusted upon claim by either party made within a reasonable time after the first observance of the conditions. [Emphasis added]

Article 16, relating to the claim for extra cost, states:

If the contractor claims that any instructions by drawings or otherwise involve extra *cost under the Contract, he shall give the Archi-*

tect written notice thereof within a reasonable time after the receipt of such instructions, and in any event before proceeding to execute the work, except in emergency endangering life or property, and *the procedure shall then be as provided for changes in the work. No claim shall be valid unless so made.*

The Architect's status and authority are related in Article 38 which is set forth in part below:

ARCHITECT'S STATUS: ARCHITECT'S SUPERVISION

The Architect shall be the Owner's representative during the construction period. . . . He *shall have authority to act on behalf of the Owner only to the extent expressly provided in the Contract Documents or otherwise in writing,* which shall be shown to the Contractor." [Emphasis added]

The contractor's primary contention was that he was entitled to more pay for work not under the contract, since it is customary for a contractor to comply with the request of the owner's representative, who in this case was the structural engineer, an employee of the bank's architect. The bank maintained that the very purpose of the contract was to assure it the building desired at the contract price of $679,560.

The court said:

We are of the opinion that the written contract expresses the agreement of the parties and that it prevails over custom. Courts do not have the power to make contracts where none exist, nor to modify, add to, or subtract from the terms of one in existence. . . .

With the above premise in mind and the contract before us, there remains the question of whether the terms of the contract between the parties permit an award on a *quantum meruit* basis. We note immediately that such an award, if proper, would require a finding by the court that the labor was not anticipated by the contract, and also that there were no provisions of the contract by which payment could be made for unanticipated labor. In the recent case of *Delta Construction Co. of Jackson vs. City of Jackson,* 198 So. 2d 592,600 (Miss. 1967), though relating to a public contract, nevertheless, expresses a principle of law we think equally applicable to private contracts. We there stated:

"The contract in the instant case requiring a supplemental agreement for extra work over minor changes is essential, because municipalities and other governmental agencies obtain funds with which to build public improvements from bond issues based upon estimates furnished to them, and municipalities must reserve the right to stop a project if they determine the extra work will exceed the amount of money allocated to any given phase of a project. . . . Furthermore, it has been generally held that no recovery can be had on an implied contract, or quasi-contract, or upon *quantum meruit* for extra work where the claim is based on an expressed contract. . . ."

The old foundations were anticipated since Section 2-F of the contract refers to them with provision of method of procedure upon their being encountered and with the further provision that no separate payment would be made for this labor. Article 15 sets forth the method of payment for extra work, notable of which is that, "no extra work or change shall be made unless in pursuance of a written order from the owner," and Article 16 provides that in any event no claim for extra work shall be valid unless the contractor gives the architect written notice of his claim "before proceeding to execute the work. . . ." Finally Article 38 limits the architect's authority on behalf of the owner to the express terms of the contract or otherwise in writing.

We can only conclude in comparing these plain terms to the vague assumption of the contractor that custom of the trade would implement the written document in his behalf, that the former prevails. The written contract anticipated every contingency upon which this suit is based. Its very purpose was to forestall imposition of vague claims derivative of custom within the trade with which laymen are often unfamiliar. The owner, being desirous of limiting its financial obligation, should not have its pocketbook exposed to the custom of architects and contractors unless it agrees thereto. In this instance the owner agreed to pay for extra work only if it was authorized in writing prior to its execution. Having contracted directly upon the point, there was no leeway for an award on a *quantum meruit* basis.

Reversed and remanded. (*Citizens National Bank of Meridian vs. L. L. Glasscock Inc.*, 243 So. 2d 67.)

The importance of anticipating every possible contingency in writing when drawing up a construction contract is clearly shown in this opinion. The successful owner, on the one hand, managed to get everything foreseeable into the contract even if it meant making such a catchall device as the entire contents of an American Institute of Architects bulletin a part of it. On the other hand, the contractor missed his opportunity to insist that change orders be made in writing as provided for in the contract, and so exposed himself to confusing instructions and a change of mind by the architect, on whom he had relied to intercede with the owner. A few words in writing would have solved the problem.

When a Vague Contract Fails, Who Assigns Value to Work Completed?

One of the first things a law student learns is that the meeting of the minds of the parties is the essence of a valid contract. In fact, one of the chief reasons for preparing detailed plans and specifications is to make certain that the minds of the parties have met on every phase of the job to be done. But what if work on a job proceeds, even though it can be proved that the minds of the parties really never met? The party who performed the work under a substantial assumption that he would be paid is permitted to recover on a *quantum meruit* basis; that is, he must prove the value of the work he has performed.

The case summarized here is an excellent example of a transaction in which the court permitted such a recovery and also discussed in considerable detail the methods by which the value of the work done by an architect and an engineer he employed might be proved. A verdict for $21,558.56 in favor of the architect was reduced on appeal by more than $12,000 because he failed to provide satisfactory proof of the value of the work he had done. The Supreme Court of Wisconsin made the ruling. Following are the facts of the case as stated in the court record.

In late 1965, a food processor and distributor, Jewett & Sherman Co., decided to build a new plant and warehouse. It contracted, among other architects, Charles H. Harper & Associates, the eventual plaintiff. Charles Harper, president of the firm, explained to Jewett & Sherman what services he could offer and how his fee would be computed should he be hired.

On July 12, 1966, Ralph Gardner, the chairman of the board of Jewett & Sherman, telephoned Harper. When asked at the trial what Gardner had said, Harper replied: "Mr. Gardner said that we had received—he didn't say, 'you are hired.' He said the building committee wanted us to do the work and we should come down and get started to do the work."

From July 12 to Sept. 22, 1966, Harper and the other employees of his firm, including architects and draftsmen, devoted time to the proposed building. Harper also asked three consulting firms to help determine the defendant's building needs. The firms were Ring & DuChateau Inc. (plumbing and heating, air conditioning, and ventilating), Leedy & Petzold (electrical), and R. C. Greaves & Associates (landscape architecture).

On Aug. 10, 1966, Harper tendered the American Institute of Architects' Standard Form of Agreement Between Owner and Architect to the defendant. The contract called for Harper's firm to be paid five percent of the construction costs. Neither party signed the contract.

During most of August and until Sept. 22, 1966, Harper was also working on other jobs in the Milwaukee area. These other jobs were, by Harper's own admission, substantial projects, one of which had a construction cost of between $4 million and $5 million.

Robert Jenstead was, during this period, the corporate engineer for Jewett & Sherman Co. and also chief liaison with Harper. Jewett was eager to get the project moving as quickly as possible. Jenstead testified that during August he was calling Harper daily and that initially he was able to contact Harper fairly easily, but as the month wore on, he was very rarely able to get in touch with him. He also stated that despite specific requests, his calls were seldom returned.

On Sept. 22, Jewett & Sherman contacted Harper and advised him to stop all work on the building. Subsequently, Jewett hired Link Builders Inc., to handle the project. Harper was told that he would be paid for his services up to Sept. 22, 1966, and that he should submit a bill to Jewett & Sherman Co. That bill was in the amount of $25,875. Harper derived that figure in this manner:

He estimated the total cost to be $1,250,000. Five percent of that figure would be $62,500; that is, if the project were completed, he would be entitled to $62,500. He then estimated that the work he had done up to Sept. 22 amounted to 35 percent of all the work he would have done had he followed through to completion. Taking 35 percent of 5 percent of the total (i.e., 35 percent of $62,500), he arrived at $21,875 as his bill.

Jewett & Sherman refused to pay the bill on the ground that it was improper to compute his bill on the basis of a contract neither party had signed. Jewett & Sherman then sent Harper a letter, by their attorney, stating that they would pay nothing until Harper submitted a statement of the time he and his associates put in on the project.

In August, 1967, Harper's attorney told him to compute the time he and his staff spent on the Jewett & Sherman job. Except for Harper himself, everyone in his office kept track in writing of how much time they were spending on each job. For an estimation of his own hours spent on the Jewett & Sherman job, Harper had to refer to his appointment book and desk calendar pad.

On Nov. 2, 1967, Harper filed a complaint that alleged a cause of action based on a contract, or, in the alternative, based on *quantum meruit*. Harper consented to a directed verdict on the cause of action for contract, and the case went to the jury on the basis of *quantum meruit*. The jury returned a special verdict in the sum of $19,448.50. Motion for a new trial on the grounds that the verdict was excessive and not supported by the evidence was denied, and judgment with costs and interest was entered in the sum of $21,558.56.

The supreme court listed the various items of Harper's bill. It began with a charge of $14,141 for architectural services, which in turn was broken down to show the work done by individual employees. Harper contended at the trial that the reasonable value of the architectural services rendered by himself and his staff could be arrived at in either of two ways. The first would be percentage computations from the terms of the contract. The second would be to accept the evidence he offered of the time spent on the project by his office. He said at the trial that when he billed on an hourly basis, he took each employee's hourly rate, multiplied that by a profit and overhead factor of 2.5 and then multiplied the resulting figure by the number of hours worked by the employee. This method produced an amount of $10,965.63 for the work of the staff.

As to this figure, the court said:

> The jury's award for architectural services was $14,141. There is no way of knowing how the jury reached that figure. If they used the hours-expended approach, then the verdict is clearly excessive in the amount of $3,175.37 (i.e., he proved $10,965.63, but they gave him $14,141, or $3,175.37 more than he proved).
>
> On oral argument respondent's [Harper's] attorney admitted that, as to this issue, there was no specific evidence to support the extra $3,175.37. He speculated that the jury must have compromised and used part of each approach (i.c., a combination of the percentage-of-cost method and the hours-expended method.)
>
> Obviously, the respondent was only entitled to be compensated once, and since each method represents a separate and distinct means of answering the same question (i.e., what is the reasonable value of respondent's architectural services?), a combination of those methods would be improper.
>
> Respondent acknowledges this, but points out that the demand in the complaint was for $21,727.12. He then argues that the jury might have used only the percentage-of-cost-of-construction method in arriving at its special verdict of $19,448.50. He contends that the court has previously approved the percentage-cost-of-construction approach and, therefore, the court should not disturb the verdict.

At this point, the court devoted more than a page of its opinion to an analysis of a 1963 Wisconsin Supreme Court case cited by Harper, *Barnes vs. Lozoff,* 20 Wis. 644 123 N.W. 2d 543, that had approved of the percentage-of-construction-cost method. The court pointed out the difference between the Barnes case and the case before it, saying:

> In the case at bar, Harper explained at a preliminary meeting that, if hired, his fee would be five percent of the total cost of construction. Much later, on July 22, Gardner called him and said words to the effect, "Come on down and get started on the job." A month later Harper rendered the AIA contract. Gardner's phone call amounted to an offer, and Harper's subsequent part performance amounts to an acceptance of

the offer. But at this point there is no valid contract because the parties never agreed to what the consideration would be. It cannot be argued that the parties impliedly agreed to use Harper's initial five percent statement for the element of consideration because that would still leave the term of consideration too indefinite to be enforceable. This is so because at the time of the phone call, no dollars and cents figure as to the total cost of construction had yet come into existence. Without such a figure, the amount represented by five percent of the cost is an unknown. It would abuse the fundamental precepts of contract interpretation to hold that Gardner bound himself and his company to pay five percent of whatever Harper should later decide would be the actual cost of the building.

Therefore, since the oral agreement never embodied the element of consideration and since the subsequent written contract was never executed by either party, it is clear that no agreement was ever reached between Harper and Jewett & Sherman Co. as to what Harper's fee should be. For that same reason (the absence of consideration), there was never an enforceable contract at all reached by these parties. Therefore, we think the appellant should not be allowed to recover on a percentage basis in this case.

Having thus disposed of the percentage–of–cost–of–total–construction method of computing the damages recoverable, the court turned to the sufficiency of the evidence to support the verdict of the jury. At the head of the list of employees who had worked on the Jewett & Sherman job was Harper's name with the sum of $5,850 opposite it. The court stated how that figure was arrived at:

> Harper testified that he personally spent 156 hours on the Jewett & Sherman [project] and that his time was worth $37.50 an hour and that this amounted to $5,850. As mentioned earlier, Harper, unlike his employees, kept no record of what hours on what days he spent on the various projects in his office. He was accustomed to working on several projects at the same time. Harper's conclusion that he worked 156 hours on the Jewett & Sherman job was based on the appointment book and desk calendar pad which he kept during the period in question. . . . It is not clear exactly what information was recorded in the two sources, but, by Harper's own testimony, his *actual* hours were not reflected therein. Apparently (the process was not explained at trial) he would find a notation of a conference or other activity connected with the Jewett & Sherman project on a given day and then estimate how many hours that activity must have taken up. Harper testified that he had no independent recollection of what hours he worked on what days but

that nevertheless he was sure that Exhibit 12 was an accurate reflection of the time he put in.

Exhibit 12, not admitted into evidence, was a compilation of the hours Harper had worked, and Exhibit 10, which was admitted, listed, with his name at the top, the hours worked by all Harper employees.

The court then indulged in a thorough discussion of the rules of evidence concerning documents used to refresh the recollection of witnesses, including a comment on the fact that Harper had been unable to produce his appointment book and desk calendar pad, and finally reached the conclusion that Exhibit 10 had been improperly admitted into evidence saying:

> The admission of Exhibit 10 was highly prejudicial to appellant, because on that exhibit Harper's hours were depicted in exactly the same manner and format as the hours of all other employees listed on Exhibit 10. Thus the jury could easily have been misled into thinking that Harper's hours rested on the same foundation as all the other hours on that sheet (i.e., valid business records).
>
> Since the Harper portion of Exhibit 10 was erroneously admitted into evidence there is no evidence of any kind in the record which would support an award for Harper's services.
>
> The amount of work Harper performed and therefore the compensation to which he was entitled must rest upon a foundation of rational and credible evidence, not speculation and guesswork.
>
> We conclude that in the absence of the Harper portion of Exhibit 10, the evidence does not support the verdict. Consequently, the amount of the verdict must be reduced by the sum of $5,850. This amount was submitted by Harper as the reasonable value of his services based on an hourly rate times 2.5 the overhead factor.

In addition to the Harper award, Jewett & Sherman challenged the amount claimed to represent the reasonable value of the services of Richard Kirsch, one of Harper's employees during the period in question. Kirsch was a Marquette University engineering student employed for the summer in Harper's office. After the July 12 phone call from Gardner, Harper sent Kirsch to Jewett & Sherman's offices. Kirsch did most of his work at the Jewett & Sherman Co. plant. There is no question about Kirsch's actual hours. He worked 160.5 hours, and Harper said his hourly rate was $5.

Five times the 2.5 overhead factor results in $12.50 an hour. and $12.50 times 160.5 hours is $2,006.25, the amount Harper claimed at trial that he was entitled to for Kirsch's services. However, Harper made an agreement with Jewett & Sherman to charge a flat $3 an hour for Kirsch's services. When Kirsch returned to school, Harper sent Jewett & Sherman a bill for $480.

Jewett & Sherman drafted and mailed a check for that amount. Harper testified he never received that check. Jewett & Sherman's records show that the check had never been cashed.

Harper testified that he agreed to $3 an hour for Kirsch's services. On cross-examination, he said he could not remember what he paid Kirsch that summer, but he did admit that it was less than $5 per hour. Harper never rendered an adjusted statement for Kirsch's services after the original bill of $480.

"We think," the court said, "respondent is bound by the agreement of $3 per hour and his billing for Kirsch's services in the sum of $480."

Having disposed of the issue in regard to Kirsch, the court concluded that in this case the proper measure of recovery was the hourly method of measuring value of services, and that the respondent could collect nothing for Harper's individual services because of failure of proof and was entitled only to $480, not $2,006.25 for Kirsch's drafting services.

The respondent's total demand and proof submitted for architectural services was $10,965.63. The elimination of claimed value of Harper's individual services and the reduction of the value of Richard Kirsch's services to $480 results in a recovery of $3,589.38 for architectural services. Accordingly, under the evidence and the law, that portion of the verdict awarding architectural services in the sum of $14,141 must be reduced to $3,589.38. This sum, added to the damages awarded for other services, results in a total verdict of $8,896.88.

The judgment of the trial court was reversed and the case remanded for further proceedings consistent with the supreme court's opinion. The defendant appellant was allowed costs on the appeal. (*Harper, Drake & Associates, Inc. vs. Jewett & Sherman Co.*, 182 N.W. 2d 554.)

4. Engineer or Architect As Supervisor

Construction contracts sometimes name an engineer or architect as supervisor of the work during the course of its performance. The cases included in this chapter tell of some of the numerous disputes that arise in connection with the duties and responsibilities thus imposed on the engineer or architect.

Is Construction Supervision an Art?

Although this case involves the supervisory duties of an architect, its legal principles apply equally to a design engineer, as the court mentions more than once in its opinion. The United States Court of Appeals, Eighth Circuit, in this case examines the responsibility of an architect or engineer for alleged negligence in his supervisory duties.

A surety company that furnished a contractor's performance bond brought this action against an architect who had prepared the plans and specifications for an airport plaza in St. Louis, Mo., and who had agreed to supervise the work in progress. The jury at the trial awarded $15,000 in damages to the surety company, but the court set aside the verdict and awarded a new trial on the issues. The surety company appealed.

The appellate court's opinion stated the facts:

Westerhold Construction, Inc. entered into a contract with the City of St. Louis, Missouri, for the construction of a Plaza at Lambert-St. Louis Municipal Airport. Aetna [the plaintiff surety company] signed this contract as surety for Westerhold guaranteeing Westerhold's performance. The Architect, in addition to designing the plans and providing the specifications, agreed to supervise the construction and as listed in the written contract with the City, undertook "to perform professional services in the preparation of drawings and specifications and supervision of construction" and "general supervisory services and advice during the construction period." Aetna knew in undertaking to guarantee the performance of Westerhold that the Architect was the "engineer or architect in charge of work." The contract of employment of the Architect, however, does not define the terms "supervision of construction" or "general supervisory" services.

The construction project met with many difficulties and delays. The Westerhold firm was in financial difficulties from the start and was unable to complete the project without financial assistance from Aetna. Apparently Westerhold used funds from this project to pay judgments resulting from the operation of other projects. Bills on this project, therefore, were left unpaid, though that phase of the work had been completed and Westerhold paid on the Architect's certification. Aetna asserts that the architect was negligent in not making any effort at any time to ascertain to what use funds were put, even after receiving reports that subcontractors were not being paid.

Aetna also alleges substandard performance by Westerhold in carrying out the construction project which necessitated increased costs and some duplication of the work. In particular: (1) A concrete retaining wall bulged noticeably, apparently due to deficiencies in the forms used and the failure to properly tie in the forms; part of the work had to be redone. An employee of the Architect noticed the misaligned forms and testified he spoke to Westerhold about it, but neither he nor anyone else with the Architect went back to check to see whether this condition was corrected before the concrete was poured; (2) the project included a sewer ditch that was 15 to 20 feet deep at its deepest point. The contractor excavated the ditch with mechanical equipment but left it open and unfilled for a period from two to three weeks, in which time the ditch eroded or sloughed (became V-shaped). The single-strength vitrified clay pipe, called for by the specifications, was unable to withstand the added pressure caused by the increased weight, due to the widening of the ditch (by erosion and sloughing), and the matter had to be corrected by redigging the ditch and replacing the sewer with double-strength pipe. There were also deficiencies alleged in the collapse of the vertical inlets to the sewer pipe, in the actual laying of the pipe, and the failure to backfill the ditch in accordance with good construction practice. The specifications provided for the contractor Westerhold to make tests of the backfilling in the presence of the Architect. However, no such tests were required or performed, nor had the Architect given the contractor any instructions with reference to the procedure to be followed on account of the sloughing condition. Because of this, some pavement had to be replaced on account of a settlement condition; (3) in addition to the repaving necessary by reason of the settlement condition of the sewer, some pavement had to be replaced because its appearance was not in accordance with specifications. Also, as an alleged result of the Architect's failure to properly supervise, the project was delayed beyond its completion date, causing additional expense in the performance of the contract.

After this recital of the facts, the Circuit Court of Appeals referred to the trial court's decision that directed a verdict for the architect despite the jury's verdict assessing damages against him. The appellate court quoted the trial court's opinion:

"Plaintiff failed to demonstrate by substantial evidence that defendant breached any duty owed to plaintiff which would create liability on defendant." The appellee Architect contends that its duty under the contract ran only to the owner [City] and since there was no privity between Aetna and it, no cause of action was stated; and further points out that the construction job was completed in accordance with the plans and specifications.

In other words, the architect might have been held responsible by the city, but not by the surety company with which he had no contractual relations.

The appellate court continued, "This case in its present posture presents the question of whether a surety on a contractor's performance bond has a right of action for loss occasioned by an architect's negligence in supervision of the construction project. The parties for the most part rely on the same cases but place different interpretations on the holding and rationale of these cases. This is a diversity case, in a proper jurisdictional amount, and Missouri law controls."

The court surmised, "We think the Missouri law allows a recovery to a surety for loss occasioned by an architect's negligence in failing to properly supervise a construction project where the architect is obligated by the contract or agreement to supervise the construction regardless of the lack of privity."

The court then proceeded to discuss a number of cases, cited by the parties, beginning with a Missouri case involving the Westerhold company that had not been decided at the time the trial court made its ruling absolving the architect of liability. In its discussion of this particular case, the court said, "*Westerhold* held that the architect was under a legal duty to the surety to exercise ordinary care in executing the certificates called for in the construction contract. The Missouri Supreme Court in a scholarly and extensive discussion of the law of privity, and its erosion through the years, said the relaxing of the rule requiring privity '. . . should be done on a case–to–case basis, with a careful definition of the limits of liability, depending upon the differing conditions and circumstances to be found in individual cases' and that the rule requiring privity would no longer be followed blindly."

Having ruled that an architect who undertakes to supervise the progress of the work is responsible to the surety despite the lack of contractual relations between them, the court turned to an interesting discussion of the measure of responsibility of architects, engineers, and members of other professions, and a useful disquisition on expert versus lay testimony. This portion of the court's opinion reads as follows:

> We think an architect whose contractual duties include supervision of a construction project has the duty to supervise the project with reasonable diligence and care. An architect is not a guarantor or insurer but as a member of a learned and skilled profession he is under the duty to exercise the ordinary, reasonable technical skill, ability and competence that is required of an architect in a similar situation; and if by reason of a failure to use due care under the circumstances, a foreseeable injury results, liability accrues. Whether the required standard of care was exercised presents a jury question. [Citations]
>
> This question was also considered by Judge Donovan in the Minnesota case of *Peerless, supra,* who aptly noted at pp. 53–954 of 199 F. Supp.:
>
> ". . . the standards of reasonable care which apply to the conduct of architects, are the same as those applying to lawyers, doctors, engineers and like professional men engaged in furnishing skilled services for compensation . . ."; and that general negligence principles apply ". . . reasonable care, which is such care as a person of ordinary prudence would have exercised under the same or similar circumstances. . ." is imposed on an architect as related to the skill, ability and professional competence ordinarily required of a person licensed to practice that profession. [Citations]
>
> Appellee Architect also contends that Aetna failed to establish the standard of care owed by a supervising architect, failed to show a breach of any standard, failed to show any damages proximately resulting from the alleged negligence, and, therefore, failed to make a submissible case. Since this case will be re–tried no purpose would be served in elaborating on the evidence. The standard of care applicable is that of ordinary reasonable care required of a professional skilled architect under the same or similar circumstances in carrying out his technical duties in relation to the services undertaken by his agreement. This includes the knowledge and experience ordinarily required of a member of that pro-

fession and includes the performance of skills necessary in adequately coping with engineering and construction problems, which skills are ordinarily not possessed by laymen. The words in the architect's contract requiring "supervision of construction" or words of similar import are not words of art and should be accorded their ordinary and usual denotation. This should be true in the absence of evidence showing a different or more restrictive connotation. There is, however, evidence in the record of what an architect would ordinarily be required to do under the circumstances presented. While this evidence is not as extensive as might be desirable, it was sufficient to make a submissible case on negligence and damages.

The court reviewed the architect's assertion that:

. . . the word "supervision" is a word of art and that evidence must be presented of what constituted due care in the exercise of professional supervision. We do not think the common term "supervision" is used as, or understood to be, a word of art in the construction contract. If it is a word of art, the general rule requiring expert testimony to establish the standard of care applied to professions requiring knowledge, competency and technical skill would be applicable. *Morgan vs. Rosenberg,* 370 S.W.2d 685 (St.L. Ct.App. 1963) clearly held expert medical testimony is essential to establish the standard of proper skill and care required of a physician, as laymen have insufficient knowledge to pass upon that issue. [Citations]
Most of the cases stating the general rule on expert testimony to establish the standard of professional care required deal with physicians and surgeons, but the same principle is applicable to attorneys, architects and engineers and other professional men. As Prosser, *Law of Torts* (3rd Ed.) points out at 164:
"Professional men in general, and those who undertake any work calling for special skill, are required not only to exercise reasonable care in what they do, but also to possess a standard minimum of special knowledge and ability."
Prosser notes that expert testimony is necessary to establish the standard of care required of those engaged in practicing medicine, as laymen are normally incompetent to pass judgment on questions of medical science or technique, but that "Where the matter is regarded as within the common knowledge of laymen, as where the surgeon saws off the wrong leg, or there is injury to a part of the body not within the operative field, it has been held that the jury may infer negligence without the aid of any expert. . . ."
It is a matter of common knowledge that it is often difficult to

secure the services of a professional man to testify in a case involving a claim of dereliction of duty by a fellow member in the profession. This undoubtedly holds true in architectural circles as well as others. There is some evidence in the case at bar regarding the duties of an architect in supervising a construction project and in carrying out his contract duties. This evidence is not as extensive as desired but was sufficient to make a submissible case. It does appear that there are certain duties patently required of the architect that are within the common knowledge and experience of laymen serving as jurors. It requires no particular technical knowledge on the part of the jury to pass upon the failure to supervise the back-filling of the sewer ditch when specifically required under the contract, the failure to correct misaligned forms utilized in retaining and supporting a poured concrete wall, or the significance of a sewer pipe that is misaligned and crooked. The jury is competent to pass on these issues without knowledge of the professional skills and competency required of architects in the ordinary performance of their skilled duties. Questions relating to stress and strain and weight-bearing capacities of structural elements are beyond the ordinary comprehension of most laymen and the court and jury require expert enlightenment on issues of this type. It, therefore, appears that the general rule requiring expert testimony to establish a reasonable standard of professional care is necessary when issues are presented that are beyond the competency of laymen jurors, but is not necessary in passing on factual situations that the ordinary layman can readily grasp and understand.

The court reversed the decision of the trial court in favor of the architect and remanded the case for a new trial at which the foregoing principles governing testimony could be followed. (*Aetna Insurance Co. vs. Hellmuth, Obata & Kassabaum, Inc.*, 392 F.2d 472.)

Who's the Boss?

The status of some engineers under the National Labor Relations Act was the issue in an action brought by Westinghouse Electric Corp. against the National Labor Relations Board before the United States Court of Appeals, Seventh Circuit. Field engineers who sometimes perform supervisory duties were involved, and the court had to determine whether these supervisory duties made the engineers ineligible for membership in the bargaining unit.

Westinghouse's steam service department is responsible for installing and servicing the turbine generators and related equipment that the company makes. The field engineers were attached to two district offices with headquarters in Chicago. At the time of the hearings, approximately twenty field engineers were attached to the Chicago offices. Employed in these offices, but outside the unit, were two clerical employees and five supervisory employees. The latter five were the central area steam service manager, the midwest and Chicago district steam service managers, and two senior service assistants.

The field engineers involved were salaried, receiving extra pay for overtime. Most of them had a bachelor's degree in mechanical engineering; the others had become skilled through on-the-job experience. About 95 percent of the time these engineers installed or serviced steam-turbine equipment at customer locations. The jobs varied in size from large installation projects that required five or six engineers, took a year to complete, and included a work force of as many as 35 craftsmen, on the one hand, to servicing jobs that took a few hours and required only one or two engineers, on the other. Any field engineer could be selected as the "lead engineer," having overall responsibility for a project. There was also an "assistant engineer," responsible for only a certain phase of the work, or for one shift on projects where work was done in shifts. Assignments to such positions depended on the complexity of the job and the ability and availability of the engineers. Six of the field engineers were usually selected as lead engineers for large projects. Occasionally, engineers from outside Chicago were called in as lead engineers on particularly complicated projects.

Westinghouse uses two approaches to installation and servicing work: the technical supervision method and the labor contract method.

Under the technical supervision arrangement, the company supplies material and equipment at specified prices and the services of one or more engineers at an hourly rate. The customer supplies the craftsmen (sometimes referred to as casual labor) and any foremen or, on large projects, superintendents.

On a labor contract project, Westinghouse agrees, for a total fee, to perform a complete installation or service, supplying the materials, engineering service, and additional labor necessary to fulfill its contract. The casual labor is hired by Westinghouse from local unions or through a labor broker. A labor contract project is usually longer and more complicated than a technical supervision job.

The original petition for an election was filed Dec. 7, 1964. On Feb. 12, 1965, after a hearing, the regional director ordered an election in a unit among professional steam service field engineers employed at the Westinghouse Chicago district offices. Service assistants, office clerical employees, guards, supervisors (as defined in the National Labor Relations Act), and all other employees were excluded.

Westinghouse sought review, and the board remanded for further hearing on whether field engineers are supervisors. Hearings were held in April, 1965. On March 31, 1967, the board issued its decision on review, reported at 163 NLRB 96.

The gist of the decision was that field engineers without lead responsibility are professional employees and the guidance or direction they give to workmen on a project is not supervision in the statutory sense. The lead engineer on a labor contract project, however, has special duties in relation to the craftsmen employed by Westinghouse. Because of these duties, the field engineer, while engaged as a lead engineer, does have supervisory status in the statutory sense. The board rejected the Westinghouse claim that every field engineer who performed this supervisory work for some part of the year should be excluded from the unit.

Relying on its own *Great Western Sugar Co.* decision, the board decided to include in the unit and qualify as a voter each engineer who, during the preceding year, spent 50 percent or more of his working time (excluding leave or waiting time) performing nonsupervisory duties. The board specified, however, that bargaining representatives may not represent any engineer with regard to his supervisory duties.

The election was held May 26, 1967, with seventeen eligible employees voting. On a vote of 9 to 8, the regional director certified the Federation of Westinghouse Independent Salaried Union as the bargaining representative of the unit.

In the unfair labor practice proceeding that followed refusal to bargain, Westinghouse opposed summary judgment and submitted an affidavit allegedly showing changed circumstances. The board granted summary judgment and issued a bargaining order.

Westinghouse contended that the following relationships are supervisory in the statutory sense: the lead engineer's relationship with the casual labor on contract projects; the lead and every other field engineer's relationships with a customer's employees on technical supervision projects; the lead engineer's relationships with other field engineers on either type of project; the other field engineers' relationship with casual labor on contract projects; and the field engineers' relationships with bladers and generator mechanics (both Westinghouse employees) assigned to contract and technical supervision projects. Westinghouse further challenged the board's concept that a field engineer assigned to a supervisory job is not a supervisor for purposes of membership and eligibility to vote in a unit if 50 percent or more of his time during the preceding year was spent on assignments where he was not a supervisor. Westinghouse also contended that all the field engineers are managerial employees, whether or not they are supervisors.

The board agreed to Westinghouse's contention regarding the relationship of lead engineers to casual labor on contract projects. It found that lead engineers on contract projects perform special administrative duties such as purchasing certain supplies, arranging through unions or labor brokers for employment of craftsmen and foremen, establishing payroll and accounting systems, paying project bills, and, in some instances, arranging to subcontract portions of the work. Although the board considered the duties of lead engineers on these projects primarily professional, as it considered the duties of all other field engineers, the board noted that the foremen directly supervising the craftsmen would seek approval of certain decisions by the lead engineers before making their own decisions final. It also noted that union stewards had presented grievances to lead engineers and accepted their decisions. The board therefore considered the relationship to be supervisory in character.

Having agreed with Westinghouse on its first contention, the court proceeded to disagree on the other six, upholding the board's ruling in each instance.

Regarding the relationship of the field engineers with the customer's employees, the court said:

> We have no difficulty sustaining the board's decision on this issue. . . . Westinghouse argues that the customer shares these employees with Westinghouse. We think, however, that the board properly viewed the arrangement as one by which Westinghouse provides technical direction and guidance to the customer's employees in order that the customer may achieve its own objective through the efforts of its employees, supervised in the statutory sense by the customer.

On the relationship of the lead engineer to other field engineers on the same project, the court supported the board's view, noting that the lead engineer has responsibility for the planning, scheduling, and successful completion of the work. Other field engineers are normally assigned to a particular phase of the work (such as piping, electrical controls, or mechanical work) or, where work is done on more than one shift, to a particular shift.

Except for the fact that the lead engineer is given wider authority and greater responsibility on a particular project, the court said that there is no evidence that he exercises or has the type of authority over the other engineers that would make him their supervisor in the statutory sense.

"The board is not bound by their testimony that the lead engineers 'give directions' or 'give instructions' to others," the court said.

Regarding the relationship of the other field engineers to casual labor on contract projects, the court went into considerable detail in an effort to define the word *supervisor,* citing a number of situations in which the circumstances made such a definition difficult:

> These instances do suggest supervisory status, although the record does not clearly establish a uniform pattern. It also appears that there are usually foremen on the job, and that the field engineer gives his direction to the foreman rather than to the man who performs the task. Our reading of the somewhat unsatisfactory record does not lead us to

conclude that the testimony compels a finding that phase engineers are supervisors of the craftsmen in the statutory sense.

Concerning the relationship of field engineers to the bladers and generator mechanics, the court, after reviewing the testimony regarding changes in conditions, said Westinghouse was not entitled to relitigate the status of the field engineers in the unfair labor practice proceeding.

The court then took up the important issue of the validity of the board's "50 percent" formula. It noted that among six engineers who had worked as lead engineers on a labor contract job, the percentage of time spent at it by each varied from 15 to 95 percent. Three spent more than half and three less. The board concluded that those who spent less were attached primarily to the nonsupervisory engineers. It noted also that a lead engineer on a labor contract project did not have supervisory authority over other engineers. The board resolved the issue by prescribing the 50 percent formula, but stipulated that any bargaining representative could not represent any engineer on questions of his supervisory duties.

The court said that the National Labor Relations Act gives the employer a right to have supervisory employees excluded from bargaining units. But the board has a duty to employees not to construe supervisory status too broadly, because the employee who is deemed a supervisor can thus be denied employee rights which the act is intended to protect.

Westinghouse pointed out that, under the board's formula, a field engineer on an assignment where he has supervisory duties would, nevertheless, be a member of the unit and eligible to vote if more than half of his working time during the last twelve months was spent on assignments where he did not have supervisory authority. On the other hand, a field engineer assigned to a job where he does not have supervisory duties would not be a member of the unit and ineligible to vote if he had spent half of his working time in the last twelve months on assignments where he did have supervisory duties.

The court said, "Obviously the situation of the field engineer who is sometimes assigned to a job where he has supervisory authority and sometimes to jobs where he has not, is ambiguous; there can be no perfect answer. We think the board's formula in the peculiar circumstances here reasonably protects the legitimate interests of the employer and employees."

Finally, in considering the alleged managerial status of field engineers, the court said that although the act contains no exemption for "managerial" employees, they have long been excluded from coverage as a matter of board policy:

As set forth in *Illinois State Journal-Register vs. NLRB*, the fundamental tests to be used in determining whether an individual is such a "managerial" employee are: Whether the employee is "so closely related to or aligned with management as to place the employee in a position of conflict of interest between his employer on the one hand and his fellow workers on the other"; and "whether the employee is formulating, determining and effecting his employer's policies or has discretion, independent of an employer's established policy, in the performance of his duties."

While the field engineers exercised a certain amount of independent technical judgment, the court held that it was the district or area managers who effectively set and carry out management policy. Only minor on-the-job changes from original plans were permitted without approval from headquarters. The lead engineer's authority to pledge the company's credit was limited to relatively small and routine purchases.

"We find substantial evidence on the record as a whole to support the board's finding that the responsibilities of the engineers are not managerial," the court concluded.

"The order of the board is affirmed." (*Westinghouse Electric Corp. vs. National Labor Relations Board,* 424 F.2d 1151.)

Does the Engineer's Supervision Cover the Contractor's Operations?

An unprecedented high tide on Ash Wednesday of 1962 (March 7) flooded a tunnel being built under the Elizabeth River between Norfolk and Portsmouth, Va. It caused extensive damage. Subsequently, one of the contractors brought an action against another contractor and the engineering firm that had designed the tunnel.

A trial in the United States District Court for the Eastern District of Virginia resulted in a judgment of $197,454.50 against the engineers but absolved the defendant contractor of responsibility for the damage. The engineering firm appealed to the United States Court of Appeals, Fourth Circuit, and the plaintiff contractor appealed the part of the verdict that held the other contractor not liable.

The court of appeals' rulings on the responsibilities of designing and supervising engineers are significant because the engineering firm had no written contract for its work; therefore, its duties had to be determined, not by written provisions of an agreement, but by the general rules of law governing such situations.

The engineering firm was Parsons, Brinckerhoff, Quade & Douglas; the plaintiff contractor was C. W. Regan Inc.; and the defendant contractor was the Diamond Construction Co.

In 1960, the Elizabeth River Tunnel Commission contracted with several contractors to build the tunnel, about 4,300 feet long. The tunnel had three generally separate sections: the "tube" section, the "cut and cover" sections, and the "open approaches." The open approaches are simply the access roads and supporting structures leading to the portal, or tunnel mouth. The cut-and-cover sections slope through the river banks into the river. They were built by cutting a sloping shaft, putting the tunnel tubes in the shaft, then recovering the shaft.

The tube portion was built by digging a ditch under the river (to a maximum depth of about 100 feet below water), by building sections of tunnel about 300 feet long and 30 feet in diameter with each end sealed by a steel bulkhead. The sealed sections were floated into place and by filling their hollow walls with concrete, or by some other means, sunk. They were then sealed together in the ditch to form one continuous tube. Thereafter the temporary steel bulkheads at the end of each section were removed, leaving one continuous tube through which the road runs. A ventilating building or tower was built near each end about 200 feet offshore from the portal.

The defendant Parsons, the engineering firm, was employed without written contract to prepare overall plans and specifications and to act as consulting and supervising engineer for the Tunnel Commission.

The defendant Diamond, the construction company, was employed under a written contract for a base fee of $3,574,000 to build the open approaches and certain retaining walls, install a roadway through the tunnel, finish the ventilating buildings, finish the entire tunnel and approaches, and install electric and drainage systems.

As for plaintiff Regan, one of its jobs was to run electrical cables within the tunnel between the two ventilating buildings. This work ultimately required portal-to-portal access to the tunnel.

At the east, or Norfolk, portal a steel tide gate was installed. It could be raised for access and lowered to seal off possible flooding through the open cut. The ground level on the Portsmouth side was higher than the Norfolk side, and the plans called for retaining walls and approaches, to be built by Regan, above the level of any recorded or anticipated tide or flood. A tide gate was not planned for the Portsmouth side.

In November, 1961, the movable steel tide gate was in place at the Norfolk end, and the only remaining metal bulkhead was inside the tunnel, about 500 feet from the Portsmouth portal and about 300 feet offshore from the ventilating building. This metal wall was interfering with the work of Diamond and Regan. Specifically, it was preventing Regan from pulling cables through from one ventilating building to the other, and it was time for it to be removed.

Diamond sought permission in November, 1961, to cut out the steel bulkhead and to build a wooden bulkhead at the portal, between the ventilating building and the shore, where the open approach to the tunnel ended. The wooden bulkhead was supposed to keep out water. Ground water and seepage were the water sources anticipated, but a flood tide was not.

Diamond drew the plans for the wooden bulkhead. Parsons examined them and recommended changes so that the bulkhead would be structurally stronger. The plans as drawn by Diamond called for a "sandwich" of aluminum foil and layers of felt between wooden timbers and for felt packing around the edges "to provide seal against concrete." Caulking the edges of the wooden bulkhead if needed and fitting its edges against the masonry to prevent leakage were, on the uncontradicted evidence, field construction details, which were the contractor's, not the engineer's, responsibility. As approved, the wooden bulkhead was constructed and the steel bulkhead removed.

By March 7, 1962, Diamond had almost completed the basic heavy structural work. The tube section and the cut-and-cover sections had been installed; the foundations for the ventilating buildings had been built by Diamond and turned over to Regan for completion, and the tunnel was open from end to end except for the wooden barricade at the Portsmouth portal and the movable steel partition, or tide gate, at the Norfolk portal.

Plaintiff Regan was the chief contractor then on the job. Regan was finishing work, including installing electrical cables between the ventilating buildings in the tunnel. Regan, however, was several months behind the contract schedule and had not installed the retaining walls on the Portsmouth side.

On March 7, 1962, a combination of winds and high tides produced a water level in the river of 10.2 feet above mean low water, more than 3 feet above any previous recorded spring tide. Shortly before 8 a.m., flood water began pouring into the tunnel from three sources—the Norfolk end, the Portsmouth end, and the Portsmouth ventilating building. It apparently started at the Portsmouth end.

At Portsmouth portal, Parsons' engineers, after a telephone request from Regan, closed the steel tide gate as far as it would go. However, they could not close it all the way; a 4-in. water discharge line and a 2-in. air line, both belonging to Diamond, and a 1-in. water line belonging to Regan all entered the Norfolk end under the steel gate and were being used by Regan. Diamond's personnel refused to remove or cut these lines because they served air lines and pumps inside.

The result was that for more than six hours water ran from the sloping tunnel entrance under about 30 feet of pressure and through the opening (which was at least 2 in. in diameter and close to 30 feet long).

Finally, some time after 2 o'clock, after the tunnel had collected large quantities of water, Diamond's men opened up this tide gate, let all the water that filled the sloping approach into the tunnel, cut the pipes and lines and closed the steel tide gate. Testimony showed that the crack underneath the tide gate was a large enough opening to fill the tunnel completely in that period of time if it had not become obstructed by debris.

A third and probably equal source of water was the unfinished ventilating building a couple of hundred feet offshore from the wooden bulkhead
near the Portsmouth end of the tunnel. Water came into the tunnel through a
number of 2-, 3-, 6-, and 8-in. holes in the foundations of the ventilating
building. Testimony indicated that the imput through this ventilation shaft
could have been more than the 90,000 gallons per hour, which would be as
great as the imput around the wooden bulkhead. Regan was in charge of the
ventilating building, and the holes in the foundations were conduits for
Regan's cables.

Summarizing the physical situation simply, the appeals court said:

> A great quantity of water came into the tunnel and caused
> damage. The water came from three sources. The plaintiff was in partial
> control of one source—the Norfolk end. The plaintiff was in sole control
> of the second source—the open ventilating building. The plaintiff was
> partly responsible for the third source—the leaky wooden bulkhead—be-
> cause of the holes he had drilled in it for pipes and wires. The defendant
> Diamond had done the planning and the building of the wooden bulk-
> head; making it watertight was Diamond's duty. The engineer, Parsons,
> was alleged to be responsible for the leaks around the wooden bulkhead
> because of Parsons' approval of the structural soundness of the plans
> drawn and submitted by Diamond.
>
> Each of the three independent sources of water could be found
> from the evidence to have contributed about equally to the damage
> plaintiff complained of.
>
> The engineer Parsons had no written contract with anyone.
>
> The Contract Documents . . . consisted of the formal contracts
> between the Tunnel Commission on one hand and Regan and Diamond
> on the other, which contracts incorporate the Plans and Specifications
> for the tunnel and the Roads and Bridge Specifications of the Virginia
> Dept. of Highways, as modified and supplemented by the Plans and
> Specifications.
>
> In the contract documents Diamond, Regan and the Tunnel Com-
> mission agreed that Parsons, as the owner's representative, should have
> certain authority to supervise and inspect and reject so as to procure for
> the Tunnel Commission the ultimate permanent result called for by the
> contracts; but the documents do not impose upon the engineer, not a
> party to the contracts, any *duty* towards the contractors of their personal
> property. To the contrary, Section 107.13 requires the contractor to
> indemnify the engineer against claims and suits arising out of the con-
> tractor's "*operations*," negligent or otherwise; and Section 107.17 ex-
> pressly provides that the engineer, as the agent and representative of the

Commission, shall have no personal liability in carrying out any power and authority under the contract or the specifications.

Independent of the contract documents, but corroborating them, the uncontradicted testimony is that Parsons' mandate from the Tunnel Commission was to supervise, inspect and reject for the purpose of seeing to it that the *permanent construction* was built in accordance with the plans and specifications. As to temporary structures, the un-contradicted testimony is that the engineer's inspection is confined to structural soundness; that approval of the contractor's working drawings for the temporary bulkhead meant simply, "that at the time the engineer knows of no good reason for objecting thereto"; that each contractor had the duty of protecting his own work; and that fitting the wooden bulkhead against the masonry and caulking it against leakage were all field construction details which were the *contractor's*—not the engineer's —responsibility. No duty to inspect the details of temporary construction nor to protect the property of one contractor from negligence of an-other contractor was shown.

Although the court had made it clear that it intended to reverse the trial court's ruling, it reviewed the plaintiff's theories of the engineer's liability. It began by stating that there were three such theories: 1) that Diamond was a third party beneficiary of an assumed contract between Parsons and the Tun-nel Commission; 2) that Parsons breached some warranty to Diamond by not properly performing his contract with the Tunnel Commission; and 3) that the relationship of the parties gave rise to a duty on Parsons to see to it that one contractor did not by negligence damage the property of another con-tractor.

The third theory received the greatest attention:

> The third theory of action—negligent performance of a duty of inspection which arose out of the relationship between the parties—is also invalid. Parsons is charged with negligence in the approval of plans for a temporary bulkhead which four months later proved leaky. There was no evidence of any duty on the part of Parsons to specify how the bulkhead should be caulked nor how it should be fitted against the surrounding masonry work. No defect in the plans was suggested nor shown. All the evidence showed that the manner of fitting the bulkhead against the masonry and the manner of caulking to prevent leaks were field details which were the responsibility of the contractor. No damage resulted from any defect in the plan. Such damage as may have resulted arose either from improper installation or from changes in the shape and fit of the bulkhead in the four months from the time it was installed

until the time of the flood. The duty of the engineer was to obtain for the owner a tunnel according to plans and specifications does not carry with it a duty to see to it that one contractor's negligence does not damage the property of another contractor, and does not create a continuing duty of inspection as to temporary details of construction of temporary structure. This theory also is inconsistent with express provisions of the contract. . . .

It is true that engineers and architects have a duty of care in drawing plans and in carrying out duties which they have accepted. It is possible, of course, for an engineer to assume such sweeping duties of supervision and control over all details of construction that nothing else appearing he may be held to have assumed a duty to parties outside his contract. [Citations]

The court concluded that Parsons had assumed no sweeping obligation and ruled that the trial court should have granted Parsons' motion for a directed verdict dismissing the complaint against it. As to Diamond, the case was remanded for a new trial. (*C. W. Regan Inc. vs. Parsons, Brinckerhoff, Quade & Douglas,* 411 F.2d 1379.)

What Is the Extent of Supervisory Responsibility?

When an engineer or architect designs a structure and agrees to supervise work during its construction, how far does his responsibility extend? That question was brought before the Fourth District Court of Appeals of Florida when a mason, who had fallen from the second floor of a building under construction, sued the architects for negligence in failing to provide an adequate guard rail. The trial court ruled in favor of the defendant architects, Ames Bennett and John B. Marion; but the plaintiff mason, Glenn Geer, appealed.

The facts of the case and the appeals court's decision follows. (Note: the court's repeated use of the phrase "engineer–architect–consultant" in its opinion shows that the decision would have been the same if the suit had been brought against an engineer instead of an architect.)

In September, 1964, Palm Beach County entered into a contract with the architects to draw up plans and specifications for a new airport terminal building at the Palm Beach International Airport and to supervise construction of the work. After the plans and specifications had been drawn, the county let the contract for the terminal building to the Arnold Construction Co. The plaintiff, Geer, was a concrete mason employed by the construction company.

At the time of the accident, Geer was pouring a slab for the mezzanine floor of the building. He fell from the second floor of the building while walking along a wooden form on the outside of a recently poured section of the slab. The building was at the stage in which the walls had not been erected on the second floor and there were no guard rails or any other protective device.

Geer brought suit against the architects for the injuries he suffered as a result of the fall.

According to the appeals court:

The fourth amended complaint as filed by the plaintiff, leaving out those parts which are formal and superfluous, alleged . . . that on or about Sept. 21, 1964, the defendant-architects were employed by Palm Beach County, Florida, to prepare detailed construction drawings, specifications and other documents for the construction of an airplane terminal complex to be located in Palm Beach County, Florida; that the defendants and Palm Beach County entered into a contract on or about Sept. 21, 1964, which in addition to committing to the defendants the preparations of construction drawings, specifications and other documents, made it the defendants' responsibility to consult and assist Palm Beach County in the selection of a resident project representative, to make daily visits to the construction site to observe the progress of the construction, to provide detailed instructions to the resident project representative, to provide proper prosecution of the construction work and supervision of the same, to ascertain and assure that the construction work was progressing in strict accordance with the plans and specifications and the requirements of the funding and regulatory agencies, to directly supervise the field activities of the resident project representative and to maintain direct supervision over the contractors in the prosecution of their work. The contract between the architects, the contractor and Palm Beach County was attached to the complaint as an exhibit and incorporated therein by reference.

That on or about March 16, 1966, the plaintiff . . . was at all times acting within the scope of his employment, working on a floor approximately 12 ft above the ground level . . . that the defendants affirmatively undertook and assumed the responsibility "as the commission's representative at the work site, the engineer-architect-consultant shall maintain direct supervision of the field activities of the resident project representative(s), and through them he shall maintain direct supervision over the contractor(s) in the prosecution of their work. . . ." and further to "serve as the commissioners' professional representative in all phases of the work . . ." and further, to make "daily visits to the construction site during construction . . . to observe the progress of the construction work, and to provide detailed instructions to the resident project representative(s) . . . such instructions intended to provide proper prosecution of the work and supervision of the work of the appropriate contractor. Both of the above services to ascertain and assure that the construction work is progressing in strict accordance with the plans and specifications *and the requirements of the funding and regulatory agencies. . . .*" The complaint further alleges that the defendants by agreeing to the above affirmatively undertook and assumed the responsibility to direct the contractor in regard to the installation of guard rails, if they were not installed, and to insure that the construction work was proceeding *in accordance with all safety regulations promulgated and administered by the Florida Industrial Commission and all other applicable regulations and ordinances;* that the failure to direct the installation of guard rails amounted to negligence by the defendants, which negligence consisted of failing to ascertain and insure that the construction work was proceeding in accordance with the safety regulations promulgated and administered by the Florida Industrial Commission's ch. 185S-6.08; failing to make necessary inspections to ascertain that safe construction practices would be used; failing to require that guard rails be used during the construction of open floors with the knowledge that the failure to use the same would result in a dangerous condition likely to cause injury to persons such as the plaintiff; failure to enforce all applicable regulations, ordinances, including the Florida Industrial Commission's safety regulations; planning, designing and supervising the construction and installation of the hazardous condition with knowledge that the same would be used and worked on by persons such as the plaintiff; that at all times material herein the defendants affirmatively undertook and assumed the duty of directing the installation of the guard rails in accordance with the safety regulations promulgated and administered by the Florida Industrial Commission's ch. 185S-6.08; which directs the construction of the nature hereinbefore described, the floor [to be] guarded by guard rails on all open sides . . . that the facts in regard to the assumption of this duty consisted of defendants or their agents, employees or representatives observing that guard rails were not being utilized on the job site in accordance with the aforesaid regulations of the Florida Industrial Commission prior to and at the time of the accident. . . . Defendants or their agents, employees or representatives advised the contractors, the contractors' superintendents or the contractors' foremen prior to the accident that guard rails should be installed. Guard rails were not installed as directed by the defendants, and the defendants or their agents, employees or representatives knew that the guard rails had not been installed. Defendants took no further action to see that the guard rails were installed, although they knew that as a proximate result of the failure to install guard rails a dangerous condition resulted, which was likely to cause injury to persons such as the plaintiff. Final allegations were to the extent that the defendants were negligent in assuming the duty to direct the installation of guard rails and then negligently carrying out that duty; that as a result of the defendants' negligent failure to install a guard rail the plaintiff suffered a fall to the ground and as a result the plaintiff was injured in and about his body and extremities, suffered pain therefrom, incurred medical expenses in the treatment of such injuries, suffered physical handicap and his working ability was impaired; that the injuries are either permanent or continuing in nature and plaintiff will suffer such losses and impairment in the future. Plaintiff demanded judgment for damages in excess of $3,000 against the defendants and requested trial by jury.

Thereafter, according to the court, the defendants filed motion to dismiss Geer's fourth amended complaint. One of the defendants, Marion, also filed motion for summary judgment. The trial court simultaneously granted all the motions and entered final judgment in favor of both defendants.
The appeals court continued:

Although the question of an architect's liability for personal injury or death has not heretofore been determined in the State of Florida, law abounds in other jurisdictions to the effect that an architect who plans and supervises construction work as an independent contractor is under a duty to exercise ordinary care for the protection of any person who foreseeably and with reasonable certainty may be injured by his failure to do so. The following cases concern themselves with an architect's liability for negligent acts. *Montjo vs. Swift,* 1963, Cal. App. 2d 351,33 Cal. Reptr. 133 (suit by pedestrian descending a stairway in a bus depot for injuries in a fall, against the architect who planned and supervised the renovation and the reconstruction of the stairway; judgment for the defendant notwithstanding the verdict, reversed and order for a new trial affirmed). *Willner et al. vs. Woodward,* 1959, 201 Va. 104,109 S.E. 2d 132 (suit by homeowners against architect for damages allegedly due to negligent plans, negligent approval of plans of others and negligent supervision of construction, relating to the heating and air conditioning system, the plans being inadequate and not requiring ducts to be encased in concrete and not providing a forewall, the architect not having inspected the ductwork and having relied upon the design and system of the contractor; summary judgment for the defendant reversed and remanded). *Erhard vs. Hummonds,* 1960, 232 Ark. 133, 334 S.W. 2d 869. (The owner employed the architects to draw up plans for a commercial building. That was done. The owner then employed the architects to supervise the construction for an additional fee. There was no written contract between the architect and the owner. The general contractor subcontracted the excavation work to another party. The plans for excavation were in some detail in the contract. The walls had to be shored to prevent sliding and caving. The field supervisor for the architect raised questions about certain of the shoring. The architects asked the general contractor to send a new job superintendent. He arrived on a Friday and promised the shoring in question would be done on Monday. It rained over the weekend and the excavation wall softened. On Monday, as the field supervisor of the architect drove his car near the embankment wall, it caved in. Three employees of the subcontractor were killed and another injured. The architects were sued for the deaths and the injury. Judgments for the plaintiff were affirmed.)

The law applicable to architects' liability for personal injury or death was summarized by the court:

[Architects] may be liable for negligence in . . . the erection of an unsafe structure whereby anybody lawfully on the premises is injured. Possible liability for negligence resulting in personal injuries may be based on [the architects'] supervisory activities or upon defects in the plans or both. Their possible liability is not limited to the owner who employed them. Privity of contract is not a prerequisite to liability. They are under a duty to exercise such reasonable care, technical skill and ability and diligence as are ordinarily required of architects in the course of their plans, inspections and supervisions during construction for the protection of any person who foreseeably and with reasonable certainty might be injured by their failure to do so.

The court then said that under the applicable law, the allegations in the plaintiff's fourth amended complaint "are sufficient to withstand a motion to dismiss."

Having ruled in favor of the plaintiff on his first point, the appeals court took up the trial court's action in granting a summary judgment to Marion, thereby refusing to let the matter go to a jury. The appeals court said in part:

An architect has been defined as one skilled in practical architecture, whose profession is to devise the plans and ornamentation of buildings or other structures *and supervise their construction.* An architect or engineer has also been defined as one whose special business it is to design buildings, fix the thickness of their walls, the supports necessary for maintenance of them in their proper position, and to do all things in the line of his profession for the guidance of builders in the erection of buildings. Architecture is the art of building according to certain determined rules. [Citation]

Decisions of other states make it clear that an architect is not under a duty to supervise construction. [Citation] However, architects do supervise as a matter of common practice, [citation] and such supervision is properly within the scope of their professional capacities. When architects do undertake supervision of construction in addition to the preparation of plans, they generally are compensated separately or additionally, and if they perform their supervisory duties in a negligent fashion their liability therefor is separate and distinct from the liability of the party who negligently performs the building process. [Citation]

In the instant case, the plaintiff attached to the complaint the contract executed by the defendant-architects and the other parties involved in the construction of the airport.

According to the court, the contract provided that a professionally qualified representative of the architects was to make daily visits to the construction site to observe the progress of the construction work and to provide detailed instructions to the resident project representatives. These instructions were intended to provide proper prosecution of the construction work and supervision of the work of the appropriate contractor.

Both these services, the court observed, were to ascertain and assure that the construction work would proceed in strict accordance not only with the plans and specifications but also the requirements of the funding and regulatory authorities. The contract stated that it was the responsibility of the engineer-architect-consultant to find out if the contractors were properly performing the work, and if not, to notify the commission. It was also the engineer-architect-consultant's responsibility to keep the commission informed of the progress of the work, to advise of deficiencies and delays, and to provide recommendations for the correction and return to schedule.

The contract further provided that the engineer-architect-consultant advise the commission during construction and act as the commission's representative at the work site, issue the instructions of the commission to the various contractors, and prepare any change orders required. As the commission's representative at the work site, the engineer-architect-consultant was to maintain direct supervision of the field activities of the resident project representative and, through him, maintain direct supervision over the contractors in the prosecution of their work.

The court noted:

There certainly are material issues of fact as to the duties imposed upon the architects under the contract term "supervision," particularly as that term relates to the taking of positive steps to insure the safety of workmen during the construction. Determination as to whether or not the architects exercised their duties is properly within the realm of the jury.

Accordingly, the final judgment is reversed, plaintiff's fourth amended complaint is reinstated, and the case is remanded for further proceedings consistent with the views herein expressed. (*Geer vs. Bennett,* 237 So.2d 311.)

If a Contract States That the Engineer's Determinations Are Binding, Can They Be Impeached?

In this case, each party in a construction contract agreed to be bound by the determinations of an engineer identified in the contract. However, the contractor, who had undertaken to build a sewer line, contended that the engineer had erred in his computation of the amount of work actually done, and, consequently, filed suit in the Oklahoma district court. The contractor sued the owner for payment; the owner contended that both he and the contractor had agreed to be bound by the engineer's determinations and refused to pay the additional sum. The District Court of Tulsa County rendered judgment for the contractor. The owner appealed to the state supreme court.

The contract clause at issue reads: "The [quantities and prices for excavation] are estimated only and Contractor shall be paid on the basis of actual quantities of work performed. The parties hereby designate Shibley–Lane Engineering Co. as agent of each of the parties for the purpose of determining actual quantities of work performed pursuant to the contract. The actual quantities shall be measured by the Engineer during the performance of the work and his determination of quantities shall be binding on both parties hereto."

The contractor contended that the engineer's determination of the amount due him was $10,439.51 less than the amount actually due him, and the Tulsa jury awarded him that amount, plus attorney's fees.

In the arguments, the owner (as well as the contractor) cited *City of Lawton vs. Sherman Machine & Iron Works* (1938), 182 Okla. 254, 77 P.2d 567. In this case, the plaintiff, again a contractor, tried to recover balances he claimed were due him, according to the final certification of the city engineer, under two contracts with the defendant, a city, for building two portions of a water system. The contracts contained this provision:

> To prevent all disputes and litigations, it is further agreed by the parties hereto that the City Engineer shall, in all cases, determine the

amount and the quality of the work of the several kinds of work which is to be paid for under the terms of this contract; and he shall decide all questions which may arise relative to the execution of the contract on the part of the contractors; and his estimates and conclusions shall be final and conclusive.

The city alleged fraud on the part of the city engineer in making the certificate. In the case under discussion, the owner quoted the second paragraph of the supreme court's syllabus to the opinion:

> Generally where an engineer is designated in a building or construction contract as the person who shall be the arbiter of the amount and character of the work done, and the amount due the contractor under the contract, the final certificate of approval of such engineer is binding upon the parties. The approval or certificate, however, may be avoided upon a showing of *actual* fraud or of such gross errors or mistakes as to constitute *constructive* fraud. [Emphasis added]

To this argument by the owner, the supreme court said:

> In connection with the matter of the avoidance or impeachment of such a certificate, this court, in the City of Lawton case, made, or quoted with approval from other jurisdictions, statements to the effect that: the law writes into such a contractual provision a requirement that the architect or engineer must, at all times, and in respect to every matter submitted to his determination under such a provision, exercise an honest judgment and commit no such mistake as, under all the circumstances, would imply bad faith, and that his certificate must be made upon such knowledge of the subject matter as to warrant an opinion binding upon the parties to the contract. To make a certificate without a proper knowledge of the facts, which upon investigation is found to be untrue, is equivalent in law to the making of a certificate known to be false, for the result is the same—one is the product of open fraud, and the other the product of such bad faith as to operate as a fraud in law. . . .

The owner also argued, in what the court called his "second proposition," that the final determination of the engineer cannot be impeached except by "clear, cogent and convincing evidence" of actual failure on his part to exercise an honest judgment, or of an error so gross that it would imply bad faith and constitute constructive fraud. The owner contended that there was no allegation or evidence of any fraud, and that the evidence about the engineer's failure to exercise an honest judgment and concerning gross errors on his part was not "clear, cogent and convincing." Therefore, the owner said, the trial court erred in submitting that issue to the jury.

In the trial court, the jury was instructed that, among other things, "gross mistake or the failure to exercise an honest judgment is never presumed and cannot be inferred from facts consistent with honesty of purpose, but in a case of this nature, where gross mistake or failure to exercise an honest judgment is alleged, it must be proved by clear, cogent and convincing, positive and satisfactory evidence which preponderates to a degree of overcoming all opposing evidence and repelling the presumption of good faith." Neither the contractor nor the owner excepted to giving that instruction. The supreme court decided, without passing on the correctness or incorrectness of such a rule of evidence, to treat that instruction as stating a proper rule for the purposes of the case.

When the trial began, each juror received a copy of an instrument marked "Plaintiff's Exhibit 14." This was the "comparison sheet," which, as its name implies, contained comparative figures representing the contractor's and the project engineer's versions of the amount of work done by the contractor, and for what he asked to be compensated under the provisions of the contract. The comparison sheet listed 46 items divided into four groups. The first group (23 items) related to the lagoon and outfall sewer project. The second group (6 items) related to the gathering lines, and the third (6 items) to the storm-sewer portion. The fourth group (11 items) referred to the lagoon and outfall sewer lines.

The supreme court said:

> While there was some conflict in the evidence on some of the points involved there was competent evidence which would reasonably support a finding that the plaintiff [the contractor] furnished all of the items of labor and materials he claims to have furnished . . . in the quantities claimed by plaintiff. It further appears from the evidence, as to the classifications of items not contemplated by the original contract, that such additional work was done with the knowledge and at least the tacit approval of defendant. We think there was competent evidence to establish the fact that the project engineer did not correctly measure the actual quantities of the work done by the plaintiff. We are of the further opinion that this evidence, on the whole, was sufficient to satisfy the above-quoted instructions of the trial court to the jury to the effect that, before the jury could return a verdict for the plaintiff, they must' find gross mistake and, or, a failure to exercise an honest judgment on the part of the project engineer in making his computations.
>
> We hold the trial court did not err in overruling defendant's demurrers to all the evidence, or in overruling defendant's motions for a directed verdict. Nor did the trial court err in submitting to the jury the

question of whether or not the determinations by the project engineer has been impeached.

The Defendant's second proposition cannot be sustained.

Because the defendant's fourth proposition related to the matter of impeaching the engineer's determinations of quantities, whereas the third proposition did not, the supreme court considered the fourth proposition before the third. The fourth proposition was that the trial court erred in giving instruction Number Four. This instruction read:

> You are instructed that a gross difference in the amount found by the plaintiff to be due and owing under the contract between the parties to this action, with the amount awarded by the engineer, may be a fact which you can consider in ascertaining whether or not the engineer's findings should be conclusive upon the parties to the contract.

In its preliminary instruction, concerning the pleadings, the trial court told the jury that "when the parties to a contract agree that an engineer shall determine the quantities of work performed pursuant to a contract and his determination of quantities shall be binding on both parties, a court and jury are without authority to disturb the engineer's findings unless the parties seeking to avoid said engineer's finding and award prove gross mistakes on the part of said engineer as would necessarily imply bad faith or the jury finds that the engineer failed to exercise an honest judgment."

The trial court defined "gross mistake" as one "that is beyond all allowance, not to be excused, flagrant, shameful, and that which brings about a great injustice; proof of mere incompetency or mere negligence or simple negligence on the part of the engineer as to the quantities of materials furnished and labor performed is insufficient to prove gross mistake or the failure to exercise an honest judgment."

The trial court further instructed the jury that before the engineer could exercise an honest judgment in ascertaining the pay quantities, he must have had knowledge of the subject matter on which his findings were based, and that unless he did have such knowledge, he could not exercise honest judgment.

In another instruction, after limiting the amount of the plaintiff's recovery, the trial court said, "If, however, you fail to find that [the engineer] committed gross mistakes or failed to exercise an honest judgment, as defined in these instructions, then your verdict will be in favor of the plaintiff in the sum of $712.24," the amount tendered by the owner as that due the contractor under the engineer's determinations.

Of all this, the supreme court stated:

Particularly when viewed in the light of the other instructions given by the trial court concerning such a contractual provision and the impeachment thereof, we cannot agree with the defendant that Instruction No. 4 invited the jury to return a verdict for the plaintiff, but think that it tells the jurors that, in determining whether or not mistakes as to quantities, if any, on the part of the engineer were such gross mistakes as to imply bad faith on his part, they would be permitted to take into consideration any gross difference between the quantity claimed and proved by the plaintiff and the quantity approved by the engineer. If it was error for the trial court to give Instruction No. 4, it was harmless error. (*Antrim Lumber Co. vs. Bowline*, 460 P.2d 914.)

Despite its involved language and frequent repetition, the Oklahoma Supreme Court made it clear that even though each party agrees on an engineer and promises to abide by his determinations, the engineer must avoid "gross errors" in his computations. If it can be shown by good evidence (and usually it is simple to make the necessary precise measurements) that he made a material mistake, the agreement to accept his determinations means nothing. The engineer has a great responsibility in such cases, and he must not only exercise an honest judgment but also prove his honesty by being right.

Can Vague Contracts Be Enforced?

A Georgia engineering firm and a paper company entered into a contract that provided for the preparation by the engineers of detailed plans and specifications for a paper mill and a fourteen-mile railroad connecting the site with two trunk line railroads. After about six months of work, the engineering firm was dismissed as it alleged, "in bad faith and without cause," and thereupon brought an action against the paper company for breach of contract, seeking to recover as damages the amount it would have made if it had been allowed to continue its work, which included supervision of the construction work, until the whole job was completed. Incidentally, the work was completed under the direction of others at a total cost of about $1,250,000. The defendant paper company filed a series of demurrers to the plaintiff's complaint. These demurrers were overruled by the Civil Court of Fulton County and the defendants appealed to the Court of Appeals of Georgia, Division Nos. 1-3.

That court affirmed the ruling of the lower court in favor of the plaintiff engineering firm and in so doing made some significant statements regarding the rights of engineers who undertake the overall design of a big job from the very beginning.

The chief argument advanced by the paper company defendant—there were a number of legal technicalities put forward also—was that the contract was so vague and indefinite that the calculation of damages would be virtually impossible. The court did not agree. Its syllabus states the facts as follows:

There are two contracts in the amended petition and a summary of the allegations follows. The first (Count 1) concerned professional engineering services to be rendered by plaintiff in the planning and development of the plant site and railroad. The pertinent provisions are:

(a) To prepare detailed plans and specifications for the site preparation construction at said paper mill plant site adapted to the location layout proposed by defendant as aforesaid, including plane and topographical survey of the property, location and design of access roads and water disposal lagoons, and planning of site grading.

(b) To survey, locate the course of, and design the said 14 mile railroad upon the route selected by defendant as aforesaid, and to prepare detailed plans and specifications for the construction of said railroad.

(c) To consult with defendant in connection with advertising for bids for the construction of such plant site and railroad, and in the letting of contracts for such construction.

(d) To supervise the construction of such plant site and railroad pursuant to the plans and specifications prepared by the petitioner.

(e) To furnish and perform all professional engineering services relating to the above, including field engineering and surveying; design and layout work; and consultation with defendant, other engineers, and other employees and contractors of the defendant.

Payment was provided for at the following rates: Engineer—$6.25 per hour; Draftsman—$4.25 per hour; three man engineering survey party—$94.00 per day plus $11.75 per hour for hours in excess of 40 hours per week.

The contract was entered into on September 15, 1960, and the plaintiff entered upon its performance. Defendant "in bad faith and without cause" ordered plaintiff off the job on April 9, 1961. At that time plaintiff had completed all preliminary work, had prepared plans and specifications for construction of the plant site and railroad and had assisted defendant in advertising for and receiving bids on and the letting of a $500,000 contract for construction of the plant site. All these services had been performed in a workmanlike manner and had been accepted by the defendant. Since plaintiff's discharge, defendant had let an additional $750,000 contract for the railroad construction and the construction of both the railroad and the plant site had proceeded pursuant to plans and specifications furnished by plaintiff.

Plaintiff engineers have been paid for services actually rendered under the contract pursuant to monthly billings. Performance of further services under the contract set out above has been prevented despite plaintiff's readiness, willingness and ability to complete performance.

Damages are sought in the amount of the difference between the total compensation plaintiff would have received for completing the performance less the cost to plaintiff of doing so. The specific quantity of engineering services necessary to complete performance is alleged along with their contractual value and the cost of furnishing the services is set out.

A second contract for testing soil and concrete, entered into only a week or two before the discharge took place, also was involved in the action, but its disposition was governed by the decision reached regarding the first contract.

The court's opinion began as follows:

> Can the contract, as alleged here, withstand a demurrer attacking it on the basis of vagueness and indefiniteness? The defendant insists that uncertainty exists in four specifics, viz.: time for performance, when plaintiff's employees were to work; who would furnish the tools; and, the time for payment. On the other hand, plaintiff contends that the contract, being one for the professional employment of engineers, would of necessity be somewhat indefinite and that these contracts cannot be drawn with the exactness of ordinary trade agreements.
>
> There is a dearth of cases dealing with engineers and their professional contracts but we think an excellent analogy can be found in the more abundant architects' cases. Georgia courts have generally grouped the two and usually architects have some engineering training and often perform engineering services.

The architect's case cited by the court in support of its ruling was *Folsom vs. Summer, Locatell & Co.*, 90 Ga. App. 696, 83 S.E. 2d 855. The contract involved was a standard American Institute of Architects contract that contained the usual provision for preparation of the plans and specifications and subsequent supervision of the work.

After quoting this provision in full, the court continued:

> Plaintiff having completed its services, declared on the contract seeking the recovery of the fees as specified in it. In dealing with a vagueness and indefiniteness attack similar to that made here, Chief Judge Felton said: "The contract is for the performance of the enumerated architectural services required in the building of the motel and is definite and enforceable. It is not necessary that the contract set out each size of each room or unit, how many windows and doors each will

contain, the type of plumbing, the type of materials to be used, etc. The very purpose of this contract is for the formulation of these details by the architect. Such details will be contained in the architect's plans, drawings and specifications which, among other services, the contract provides for. If detailed plans, drawings and specifications were in existence and had to be incorporated into this contract, there would be no need for a contract with an architect to perform these very services."

We view this direct holding as controlling. Certainly the contract here is no less definite than that involved in *Folsom*. Further support may be found in *Curtis vs. Burney*, 55 Ga. App. 552 (1), 190 S.E. 866. There plaintiff was employed to draw plans and specifications and supervise construction of a house at $30.00 per week. After the plans were drawn and construction begun, defendant dispensed with plaintiff's services and completed the house under another's supervision. Plaintiff was allowed recovery of $30.00 a week for the time it actually took to complete the house after his discharge.

The court then cited supporting cases from California, Connecticut, Kentucky, New Jersey, North Dakota, Oregon, Pennsylvania, Texas, and Washington, and quoted from a leading textbook on building contracts, which stated the English common law view.

The court in discussing the situation before it, then continued:

> These contracts had sufficient elements by which the amount to be paid can be determined in that the overall scope of the project is clearly delineated and the amount to be paid for the services of the separate classifications of people is specifically set out. Furthermore, the petition alleges that the mill is substantially complete and thus the actual performance of other engineers would also serve to provide a measure. [Citation] The specifics of these matters are, however, evidentiary to nature and nothing further need be said here. The answer to the question of who would furnish the tools seems apparent. The engineers are contracting to do the job and it is their responsibility in the absence of agreement otherwise. As for time of payment, if nothing is specified in the contract, the time is at the completion of the contract. [Citations]
>
> The contracts alleged in both counts here are not too vague and indefinite to be enforced.

The remainder of the opinion is taken up with a discussion of the legal technicalities raised by certain of the demurrers, all of which were overruled. The way was thus opened for a trial of the case on its merits as sought by the plaintiff engineering firm. (*Southern Land, Timber & Pulp Corp. vs. Davis and Floyd Engineers Inc.*, 135 S.E. 2d 454.)

Who Has Final Responsibility?

Coordination of all contractors' work on a construction job is increasingly difficult. Much of the difficulty is the matter of responsibility—who does the coordinating?

A ruling by the New York Supreme Court sheds light on this question.

The case arose from the General Building Contractors of New York State's action to prevent the architect and owner of a college building project from shifting responsibility for the coordination of the work from themselves to the contractors. The day before bids were to be opened, the contractors obtained a stay from the Supreme Court (in New York, the Supreme Court is a court of first instance, not an appellate tribunal) prohibiting the opening of bids until the specifications could be reviewed by the court and the question of responsibility resolved.

The basis for the decision was a New York statute (Section 101 of the General Municipal Law) that requires separate specifications for plumbing, heating–ventilating, and electric wiring in public projects involving the expenditure of more than $50,000. The law also permits separate and independent bidding.

Justice Richard J. Cardamone, who ruled on the contractors' motion, stated the situation as follows:

> The petitioner is a New York Membership Corporation known as the General Building Contractors of New York State. It consists of 180 general contractors who perform a substantial volume of the public works building construction in this state. Some of its members are residents of Oneida County. The principal respondent, the County of Oneida, is the owner of a proposed public project, a Library–Academic Building, to be constructed at Mohawk Valley Community College in Utica, Oneida County, New York.
>
> Sealed bids on this proposed construction were received by the respondents on April 26, 1967, at which time they were to be publicly opened and read. This proceeding, instituted by the petitioner by show cause order dated April 25, 1967, contained a stay issued by this Court (Aronson, J.) prohibiting the respondents from opening any of the bids submitted on the Library–Academic Building until the return date of this show cause order. The stay has been continued pending this determination.
>
> The petitioner contends that the form of specifications prepared by the owner and the architect for this multiple contract project, dated March 29, 1967, violates the requirements set forth in Section 101 of the General Municipal Law. It claims that the language used in the specifications has the effect of shifting responsibilities required by the statute to

be imposed upon the owner and/or architect to the contractors, resulting in confusion insofar as the orderly progress of this public works project is concerned.

The respondents assert that petitioner has no standing in court since there is no showing that any of the members of the petitioner corporation are actually bidders on the Mohawk Valley Community College project and that this is not a class action nor a taxpayers' action. Respondents further assert that the specifications as prepared do not violate any statute or law of this state; nor do they impose any additional requirements on the general contractors bidding on this project. Finally, respondent argues that even if some additional work is required by the specifications such can be compensated for in the bid submitted by any contractor who desires to receive the award for this particular project.

The court first took up the question of whether the contractors' association could maintain its action. On this point it said:

The threshold question is whether the petitioner has standing to initiate a proceeding in this court. The preparation of specifications, advertising for bids and awarding contracts for a public project is a matter of public interest which relieves petitioner of the obligation to show that it is an aggrieved party or that it has any special interest. [Citations] An Article 78 proceeding, such as the one before this court, may be instituted by one who is a citizen, resident and taxpayer even though he has no personal grievance or personal interest in the outcome shown. [Citations] The petitioner has standing to institute this proceeding in this court.

Having thus declared the contractors' association had standing in his court, Justice Cardamone proceeded to set forth the language in the specifications to which the contractors objected:

The language used in the specifications which petitioner claims violates the statute is found at pages SC-9, SC-10 and Addendum No. 1 (1-2) of the specifications for this project. It is there provided that the contractor will "check shop drawings . . . to make sure they conform to the intent of drawings and specifications and for contract requirements. Correct drawings found to be inaccurate or otherwise in error. . . . The contractor will be fully responsible for the accuracy of such drawings and for conformity to the drawings and specifications, regardless of the approval of the architect, unless the contractor notifies the architect in writing of any deviations at the time he furnished such drawings (SC-9).

The general construction contractor shall be responsible for the proper fitting of all work. . . . Within 30 days of the execution and delivery of the contracts, the contractor for 'Contract No. 1—General Construction' shall submit to the architect for approval a satisfactory progress schedule covering total sequence and expected status of the work at any time involving the work for 'Contract No. 1, 2, 3 and 4.' " (SC-10) The substance of these requirements relating to shop drawings and progress schedules is repeated in the "Special Conditions" of Addendum No. 1 at page 2. Petitioner contends that it is these provisions which are an attempt to assign supervisory work to contractors which should be the responsibility of the owner and/or of its agent, the architect, and that such are a violation of Section 101 of the General Municipal Law.

The court then ruled in favor of the contractors saying:

At the root of this controversy lies the question upon whom shall devolve the day-to-day responsibility for the orderly progression and coordination of the public project. The provisions contained on Pages SC-9, SC-10 and Addendum No. 1 relative to the general contractor's responsibilities for shop drawings and progress schedules, particularly that expression which makes the general contractor "fully responsible for the fitting of all work . . ." appears to this court to impose upon the general contractor responsibilities and obligations as to supervision and coordination not envisioned by the statute and which should be borne by the owner and/or the architect.

Section 101 of the General Municipal Law provides in substance that in any project which exceeds the sum of $50,000 the owner will prepare separate specifications for the three subdivisions, plumbing, heating and ventilating, and electric wiring (subdivision 1). While the statute is silent as to whether the municipality may assign the work of such supervision to the successful bidder, it appears to this court that had the Legislature intended such assignment where separate bidding is required, it would have so provided.

The court then cited other statutes containing language similar to that of Section 101 of the General Municipal Law and quoted from a 1946 Governor's Memorandum in which it was admitted that the separate specifications and separate bidding would increase administrative problems but that this would be compensated for by lower costs on construction projects.

The court concluded:

. . . the state recognized the increase in administrative problems imposed upon it by the multiple bid contract system but accepted this burden in view of the over-all savings.

The interpretation of the statute (Gen. Mun. L Sec 101) must be read and given effect as it was written by the Legislature and not as this court thinks it might have been written if the Legislature could have imagined all of the problems which might arise under it. Here, it appears that the County is attempting to exercise powers not expressly granted it under the statute and which can be exercised only where it is "so essential to the exercise of some power expressly conferred as plainly to appear to have been within the intention of the Legislature." The implied power must be necessary, not merely convenient, and the intention of the Legislature must be free from doubt. [Citation] It does not appear that the shifting of responsibility for coordination and supervision from the County or its agents, in these cases its architects, may be shifted to any of the prime contractors. The power which the county seeks to imply into the language of the statute is not a necessary one, but merely a convenient one. (*General Building Contractors vs. County of Oneida,* 282 N.Y.S. 2d 385.)

The contractors' association had asked that new specifications be drawn, but the court ruled that the desired result could be obtained by eliminating from the specifications the objectionable language. Although the multiple-contract system is not used in many jurisdictions, this decision is of interest because of the recognition of the right of the contractors' association to intervene even though none of its members was an actual bidder on the job at stake, and the successful result of the association's timely effort to prevent a shifting of supervisory responsibility from owner to contractor.

Where Do Supervision Responsibilities End?

What are the duties and consequent responsibilities of the engineer or architect who not only prepares plans and specifications for a construction job but also agrees to supervise the work as it progresses? The Court of Appeals of Arizona, an intermediate appellate court, offered valuable answers to that question in deciding a case against the owner and the architect of a school building brought by three employees of the general contractor who were injured when some steel work collapsed. (Although the case involved an architect, the legal principles would be the same if an engineer had designed and undertaken to supervise the job.)

The trial court rendered judgment in favor of the defendants, but the plaintiff appealed the decision. The court of appeals brought out the following facts.

In December, 1962, the Board of Supervisors of Maricopa County, Arizona, acting on behalf of the defendant school district, entered into a written agreement for architectural services with the defendant, a registered architect. The agreement provided for the preparation of plans and specifications for construction of a physical–education building, including a gymnasium, and general supervision of the related work. The project was presented for bids; Verdex Steel & Construction Co. was selected as general contractor.

Verdex fabricated and erected the structural steel used in the project. The shop drawings relative to the steel—its size, dimension, etc.—were prepared by Verdex and approved by the architect. The architect's plans and specifications did not spell out a method or sequence of steel erection but referred to the "Specifications for Design, Fabrication and Erection of Structural Steel for Building," published by the American Institute of Steel Construction, which left the method and sequence of erection to the discretion of the contractor.

The architect designed the roof of the gymnasium to be supported by six three–hinged steel arches. On Oct. 15, 1963, the general contractor's workmen, including the plaintiff, a structural iron worker, unloaded unassembled portions of at least three of these arches and bolted them together. Two cranes were used to lift the assembled parts of the arches so that they could be fastened to the vertical support columns already in place. Three arches were erected that day without incident. On Oct. 17, the plaintiff was connecting additional steel purlins to the arches when the structure collapsed, seriously injuring him.

The appeals court noted:

> In addition to urging negligent supervision as a basis for imposing liability on the owner and the architect plaintiff contends that both should be held strictly liable in tort for alleged defects in the plans and specifications drawn and submitted by the architect. However, we need not decide whether the strict tort liability doctrine as applied to product liability cases is applicable to the "manufacturer" or designer of buildings, because in this case, there was no evidence which would support a finding that the plans and specifications were in fact defective. Plaintiff's briefs do not point out any such evidence, nor has the court discovered any in our reading of the record.

The plaintiff's primary contention was that the evidence supported a finding of liability based on the negligent exercise of supervision over the work of the general contractor retained by the owner and his representative, the architect.

The question of liability for injuries to employees of an independent contractor where the owner or other employer of that contractor has retained certain supervisory powers over the work and has vested these retained powers in a representative or other employer has been presented many times to the Arizona appellate courts. All cases support the principle that liability for negligent exercise of retained supervisory powers can attach only when it can be shown that a duty has been created by the stipulation of "the right to exercise day-by-day control over the manner in which the details of the work are performed."

The appeals court said:

> As stated in *German vs. Mountain States Telephone & Telegraph Co., supra,* the retained supervisory controls must give the defendant control over the method or manner of doing the details of the work over and above the supervision and inspection rights generally reserved to make certain that the results obtained conform to the specifications and requirements of the construction contract.
>
> Although none of the [cited] Arizona decisions involved the vesting of these retained supervisory powers in an architect, we believe that, insofar as concerns the question of whether or not the contract documents have created a duty to the injured employee of an independent contractor, the principles developed are analogous and applicable to a fact situation involving an architect. If it is found that a duty has been created so that the architect has the duty to supervise the method and manner of actually doing the work, then the fact that the person in whom such supervision is vested is an architect might be material in ascertaining whether that supervision has been negligently exercised. In other words, assuming the existence of a duty, a different and more exacting standard of conduct might be imposed upon an architect than would be imposed upon an unskilled owner retaining such supervisory rights.
>
> We are not unmindful of cases from other jurisdictions relied upon by plaintiffs which impose liability upon the architect or other owner's representative without finding the existence of such a duty.

The court then cited the case of *Miller vs. DeWitt,* 37 Ill. 2d 273, 226 N.E. 2d 630 (1967), noting that other cases have disregarded fundamental contractual principles in attempting to parlay general inspection or supervision clauses that give the owner or architect a *right* to stop observed unsafe construction processes into a *duty* which is neither consistent with generally accepted use contemplated by the contract or the parties. The court cited the dissenting opinion in *Miller, supra,* as constituting, in its view, a concise, well-reasoned statement of the controlling legal principles, and consonant with the Arizona decisions cited.

The court then turned to the contract documents involved in the appeal—i.e., the contract between the defendant architect and the owner, and the contract documents between the owner and the general contractor. First, the defendant architect, citing *Blecick vs. School District No. 18 of Cochise*

County, 2 Ariz. App. 115, 406 P, 2d 750 (1965), held that the extent of his duties should be determined solely by the provisions of his contract with the owner and that since he was not a party to the contract between the owner and the general contractor, the provisions of that contract would have no relevance.

The court commented:

> As a general proposition we would be inclined to agree that the contractual obligations of a party must be measured by the provisions of that party's contract, and that such obligations could not be enlarged by a separate independent contract without that party's consent. However, we do not believe that this principle is necessarily applicable under the facts of this case. Here, the evidence shows that not only did the architect prepare the contract between himself and the owner, but, in addition, he subsequently prepared the contract between the owner and the general contractor. Under these circumstances, it is our opinion that if the provisions in the contract between the architect and the owner relating to the architect's supervision operations are ambiguous, then the provisions in the contract between the owner and the general contractor might become relevant to resolve such ambiguity.

In examining the provisions of the contract between the owner and the architect, the court found limited reference to the supervisory obligations vested in the architect. Provision was made for payment to the architect of 2 percent of the total cost "for supervision of the work." Under paragraph 1, entitled "The Architect's Services," it was provided that, among other duties, the architect's professional services included "Supervision of the Work." Paragraph 5, "Supervision of the Work," provided for "general supervision to protect the owner against defects and deficiencies in the work of contractors," but the architect did not guarantee the performance of their contracts. This "general supervision" was distinguished from "continuous on–site inspection by a clerk-of-the-works."

The appeals court said:

> In our opinion provisions such as these are insufficient to support, and in fact negate, a finding that the parties intended that the architect have the obligation or duty to control the contractor and his employees in the method and manner of doing the details of the work. Such provisions are more consonant with an intention that the architect have such supervisory controls as are necessary to assure that the results of the contractor's work comply in technical detail with the plans and specifications prepared by the architect. From the evidence it is clear that the parties to this contract so construed its provisions.

Ordinarily, this fact alone would be dispositive as to what the parties intended. However, inasmuch as the liability of these parties depends upon the extent of supervisory rights retained by the owner and placed in his representative, the architect, we would be inclined in this case not to give too much weight to their oral expressions of what was intended if these expressions were in any way in conflict with the provisions of the contract. Here we find no such conflict with the terms of the architect's contract.

Even if we were to consider that the provisions of the contract between the owner and the general contractor could somehow serve to enlarge the architect's supervisory duties, and further, assuming that plaintiff was injured because of an unsafe steel erection procedure adopted by the general contractor, reference to that contract would not aid plaintiff. A full reading of the documents constituting the contract between the owner and the general contractor shows clearly that the method and manner of steel erection . . . is left entirely to the discretion of the general contractor without any retained right to supervise or control such procedure. Sec. 9.03, subsection A of the contract's specifications reads:

"Materials. The structural steel shall strictly conform to the applicable requirements of the specifications for design, fabrication and erection of structural steel for building, published by the American Institute of Steel Construction."

The applicable provisions of said publication, made a part of the contract by said reference and admitted in evidence, provide as follows:

"Sec. 7. Erection (a) Method of Erection. If the owner wishes to control the method and sequence of erection, he so specifies in the invitation to bid or the specifications which accompany it. Otherwise the fabricator will proceed according to the most economical method and sequence available to him consistent with the plans and specifications and such information as may be furnished to him prior to the execution of the contract."

From these provisions, it is clear that the owner did not by contract reserve any right in itself or impose any duty upon itself or the architect to control the steel erection procedure which resulted in plaintiff's injuries.

Further, contrary to plaintiff's contentions, it is our opinion that A.R.S. Sec. 32-142 (Arizona statutes), subsec. A, does not increase the duties assumed by the architect in his contract with the owner.

A.R.S. Sec. 32-142, subsec. A, states that drawings, plans, specifications and estimates for public works of the state or a political subdivision thereof involving architecture, engineering, shall be prepared by or under the personal supervision of, and the construction of such works shall be executed under the direct supervision of, a registered architect or engineer.

The plaintiff held that this statute imposed on the architect the duty of directly supervising the procedures and techniques used by the general contractor in the construction of public works and that the duty was imposed to protect the safety of the general contractor's employees at the construction site. If such supervisory duty was created to protect persons in the plaintiff's class, the court declared, it would follow that negligent exercise of that duty would allow legal action on the part of a worker injured as a result of a breach of that duty.

The court said that while there has been no case defining "supervision" as used in A.R.S. Sec. 32-142, subsec. A, some light on the intended meaning may be found in the definition of "architect" used in connection with the adoption of the Technical Registration Act of 1935, Chapter 32 (1935), laws of Arizona, the forerunner to A.R.S. Sec. 32-142, subsec. A. In this act, *architect* was defined as "a person other than an engineer who prepares drawings or specifications or supervises *but does not superintend* the construction of buildings and structures, as an authorized agent of the owner thereof." [Emphasis added]

The court admitted that the Technical Registration Act did not control the present case but was helpful in determining the originally intended scope of the supervision function. The court stated:

> This original legislation coupled with the present provisions of Sec. 32-101, subsec. 2, which defines "architectural practice" as including the "supervision of construction for the purpose of assuring compliance with specifications and design," makes it clear that the supervision provisions of A.R.S. Sec. 32-142, subsec. A, were intended solely to assure that the results obtained comply with the construction specifications and design. The type of supervision contemplated therein is not a detailed supervision of construction. These statutory provisions are much too general to support the duty plaintiff would have us find.
>
> We hold that neither the contract documents nor the statute imposes any duty upon the architect or the owner to supervise the procedures to be utilized by the general contractor in the temporary erection of the steel structures which collapsed causing the plaintiff's injury.

In its concluding paragraphs, the court's opinion rejected minor points urged by the plaintiff and ruled that the decision of the trial court absolving the architect and the owner of all responsibility for the collapse of the steel arches be affirmed. (*Reber vs. Chandler High School District*, No. 202,474 P 2d 852.)

5. Engineer vs. Architect

The work of the engineer and the architect often overlap, and their overlapping has been the genesis of many lawsuits. The cases in this section demonstrate how some courts have attempted to resolve such conflicts. It should be noted, however, that so far as the basic legal principles are concerned, the conduct of the engineer and of the architect is usually governed by the same rules. Courts like to reason by analogy. The analogy between the engineer and the architect is a favorite one.

Can the Architect Charge for Supervisory Services?

A California court, the Second District Court of Appeal, Div. 2, has made a contribution to the discussion of the relative functions of architects and engineers. The action involved a dispute between an architect and the owner of a building constructed from plans made by the architect who also supervised the actual construction. In defining the architect's functions, the court commented on the services of engineers involved in comparable activities.

The owner of the building didn't want to pay the architect for his supervisory services, contending that the architect could not collect for them because he was not a licensed contractor and that the supervision of the job was outside of his role as a duly licensed architect. In reaching a decision in favor of the architect, the court discussed the comparable role of engineers engaged on similar jobs, saying:

> The entire structure of appellant's arguments on appeal rests upon a fundamental assumption that an architect is "acting solely in his professional capacity" only when he is engaged in designing buildings and drawing the plans therefor; and that whenever he steps beyond the strict limits of this role and undertakes to supervise the construction of a

building as an owner's agent, he no longer is acting in his professional capacity as an architect, and must, if he is to be compensated for such services, possess a contractor's license also.

Neither reason nor the applicable authorities support this assumption. Although there is no statutory definition of an architect's "professional capacity," a professional engineer, who is similarly exempted by section 7051 is defined in section 6701 as:

"[A] person engaged in professional practice of rendering service or creative work requiring education, training and experience in engineering sciences and the application of special knowledge of the mathematical, physical and engineering sciences in such professional or creative work as consultation, investigation, evaluation, planning or design of public or private utilities, structures, machines, processes, circuits, buildings, equipment or projects *and supervision of construction for the purpose of securing compliance with specifications and design for any such work.*

"Any person practices civil engineering when he professes to be a civil engineer or is in responsible charge of civil engineering work." (Section 6734.)

"The phrase 'responsible charge of work' means the independent control and direction, by use of initiative, skill and independent judgment, of the investigation or design of professional engineering work *or the supervision of such projects.*" (Section 6703.)

It would seem to present an extremely anomalous situation if *engineers* were held to be exempt from obtaining contractor's licenses when performing supervisory services, while *architects* performing the same services were not. This is particularly apparent when it is noted that Section 6737 provides that "an architect who holds a certificate to practice architecture in this State . . . insofar as he practices architecture in its various branches, is exempt from registration under the provisions of this chapter [relating to engineers]." (*Wallach vs. Salkin,* 33 Cal. Rptr. 125.)

In another case, a court found that in the State of Washington an engineer may perform architectural services if he is careful not to present himself to the public as an architect. The decision was handed down by the Supreme Court of Washington, Department 2, which interpreted the state's statutes as stated above and permitted an engineer to maintain an action for architectural services.

There was no question in regard to the nature of the services rendered by the plaintiffs who brought the action. They were architectural. The supreme court's opinion relates the facts as follows:

Plaintiff Strickland was a professional engineer licensed by the state. Plaintiff Frey, although he had been an architect in Holland and was employed by an architectural firm in Seattle, was not licensed to practice architecture in the State of Washington. There is no claim that either plaintiff represented himself to be a licensed architect; in fact the defendant was informed that the plaintiff was not a licensed architect prior to her first meeting with him. . . .

In October 1959, plaintiff Strickland discussed with the defendant her need of professional help in planning and designing an addition to the Kent City Nursing Home.

October 24, 1959, both plaintiffs met with the defendant and entered into an oral contract whereby the plaintiffs were to perform architectural and engineering services pertaining to the proposed addition to the nursing home. Plaintiffs were to be paid six per centum of the total cost of construction.

Immediately, plaintiffs commenced to perform their contractual architectural and engineering services, which included: numerous conferences, preliminary studies, preparation of preliminary plans and ward plans, preliminary and final working drawings including architectural and engineering-structural, mechanical and electrical drawings, construction of a scale model of the proposed addition to the nursing home, submission of blueprints to the State Health Department and the county engineers, corrections and revisions as required by the State Health Department together with a 139 page book of specifications.

May 23, 1960, defendant became dissatisfied and notified plaintiffs that their services were terminated.

Where Does Engineering End and Architecture Begin?

An engineer who had prepared plans and specifications for a Florida shopping center was accused of practicing architecture by the Florida State Board of Architecture. The matter was taken to court. The trial court found in favor of the board and enjoined the engineer from further practice of architecture. He appealed the ruling to the Fourth District Court of Appeal of Florida. The court then had to answer the puzzling question of where the practice of engineering ends and architecture begins and vice versa.

Unfortunately, the appellate court's decision was not clear cut. It ruled in favor of the engineer, reversing the judgment of the trial court, but did so by construing two apparently contradictory Florida statutes, rather than acting in accordance with the broader principles involved. The court did include in its opinion, however, a discussion of some of the principles that govern the activities of the two professions, and which, in all probability, influenced the court's decision in favor of the engineer.

That decision seems to say that, under the Florida statutes regulating the practice of engineering and of architecture, members of each profession may practice the other profession when it is merely *incidental* to the practice of their own profession. But the decision does not give clear indication of what is considered incidental.

The Florida State Board of Architecture filed a complaint against appellant Alex Verich, a registered professional engineer, claiming that he was practicing architecture. He was enjoined from further practice; the appellate court reversed the judgment. The court noted that Verich was registered as a professional engineer in the State of Florida as provided under Chapter 471, F.S.; that he had prepared plans for a shopping center in West Palm Beach, Fla.; and that these plans involved both architecture and engineering. Verich did not hold himself to be an architect in any way, nor did he enter into any contracts to perform architectural services. He prepared the plans for the Edward J. DeBartolo Corp. for whom he was a full-time employee. (The Palm Beach Mall is owned by Palm Beach Mall Inc., a corporation wholly owned by the Edward J. DeBartolo Corp.)

The architecture board called as witnesses three architects who testified that in their collective opinion the drawings, plans and specifications for the Palm Beach Mall constituted the practice of architecture. Several registered engineers appeared for the defense and said that preparation of the plans was the type of work done by engineers. They further testified that in their opinion the engineering services rendered in preparing the necessary drawings and plans constituted approximately 75 percent of the work and that architecture was approximately 25 percent of the work. Both engineers and architects agreed that the Palm Beach Mall was a building and that the plans required the use of mathematics and the principles of engineering.

The appellate court mentioned the conflicts of opinion between the architects and engineers who testified but said its decision did not turn on a factual basis. The court found:

> That there is an overlapping between the two professions appears to be readily recognized by the language of F.S. Section 467-09(1) (a), F.S.A. which provides in substance that registered professional engineers are not prohibited from performing architectural services which are purely incidental to their engineering practice, and that registered architects are not prohibited from performing engineering services which are purely incidental to their architectural practice. The overlapping nature of the two professions is well recognized.

The court pointed out that Florida defines the practice of architecture "in a somewhat negative manner." After first providing for certain exceptions, the court said, the statute defines the practice of architecture as activity carried on by "any person who shall be engaged in the planning or design for the erection, enlargement or alteration of buildings for others or furnishing architectural supervision of the construction [and such a person shall] be required to secure a certificate and all annual renewals thereof required by the laws of this state as a condition precedent to his so doing."

The court observed a contradiction in this language by saying that the design for and erection of buildings for others is the practice of architecture, whereas the same statute, which expressly provides that "no professional engineer shall practice architecture or use the designation 'architect,' or any term derived therefrom, at the same time also expressly provides that nothing in the state law shall be held to prevent a registered professional engineer (or their employees or subordinates . . .) from performing architectural services which are purely incidental to their engineering practice."

The court said that the sense of Chapter 471 F.S. (Section 471.02 [5]) defines "professional engineering" to include any professional service requiring use or knowledge of mathematics and the principles of engineering rendered or offered for public or private buildings and any consultation, investigation, plan, design, or responsible supervision of construction in any public or private buildings.

The court said, "Thus, it can be seen that the preparations of plans and design for and responsible supervision of the construction of buildings is by statute defined as the practice of engineering."

The court then noted a paradox:

> If the planning and design of a building and the furnishing of supervision of its construction are functions which are encompassed solely within the practice of architecture, then professional engineers are prohibited from engaging in such functions unless incidental to their engineering practice. But paradoxically, the practice of professional engineering expressly includes the planning and design of buildings and the supervision of their construction. Thus, the apparent conflict can only be resolved by concluding that the statutes mean a registered architect can plan and design and supervise construction of a building as the practice of architecture and a registered professional engineer can plan and design and supervise construction of building as a professional engineer. Of course, the professional engineer cannot represent himself as being an architect nor can the architect represent himself as being a professional engineer.

The Florida State Board of Architecture contended that the statutory definition of "professional engineering" has reference to buildings of an industrial nature designed primarily to house machinery and equipment rather than designed primarily for occupancy by humans. The appellate court admitted that such a qualification would help to preserve a line of demarcation between the two professions consistent with generally accepted concepts. The court went on to say, however, that "had it been the legislative intent to thus limit or restrict to industrial buildings the type of buildings which professional engineers were authorized to design and supervise construction upon, such qualifications or restrictions would have been explicitly contained in the statute. We are concerned here with statutory construction and not historical differences of these two professions."

The judgment was reversed in favor of engineer Verich. One Justice dissented but wrote no opinion in support of his views. (*Verich vs. Florida State Board of Architecture,* 239 So. 2d 29.)

Who Is an "Engineer"?

Although it did not finally decide the issue, but sent the case back for a new trial, a New Jersey appellate court (the Appellate Division of the Superior Court) made some intriguing comments on the question, "Just what is a professional engineer?"

The case before the court was brought by an engineering firm to collect money under the terms of a contract. The defendants refused payment on the ground that the plaintiff had not procured an engineering license as demanded by the New Jersey statutes. The trial court granted summary judgment for the defendant, and the plaintiff appealed to the Appellate Division. The Appellate Division reversed the judgment of the lower court, sending the case back for a new trial, and in so doing made the following observations concerning the status of engineers.

"Engineering" today includes a vast number of subdivisions and specialties which have nothing to do with the design and structure of buildings. Examples which immediately come to mind are marine, aeronautical, electronic, computer, space, rocket, mining, radar, telephonic, television, acoustical, traffic, sound, radio, weather, missile, soil, sanitary, hydraulic, chemical, automotive, illuminating, heating, air conditioning, packaging and refrigeration. In addition, we have those who call themselves industrial, sales, efficiency, time–study, management and similar "engineers."

On the other hand, does one require a license if he does not hold himself out as a "professional engineer," but merely as an expert? Compare what H. L. Mencken said in *The American Language,* p. 291: "next to engineer, expert seems to be the favorite talisman of Americans eager to augment their estate and dignity in the world. Very often it is hitched to an explanatory prefix, e.g., housing-, planning-, hog-, erosion-, marketing-, boll-weevil-, or sheep-dip. . . ."

And what about inventors, developers of items for sale, etc.? Must all such people obtain licenses. . . ? What is meant by "professional engineering," the only occupation licensed by the Board? The statute lumps engineers with surveyors. Is it pointed only to those who deal principally with buildings, structures, and machinery equipment and fixtures? Without hearing from the Board we should not attempt to define the limits of the statute.

The court, as just indicated, expressed the wish that the State Board of Professional Engineers and Land Surveyors be heard at the new trial. (*Magasiny vs. Precision Specialties Inc.,* 221 A.2d 755.) South Carolina and Connecticut also have made contributions to settling the legal controversy concerning precise definition of the practice of architecture. One South Carolina case concerned an "architectural designer" who seemed to be stretching that nebulous term to the limit, and a Connecticut case discussed the activities of a carpenter and part-time student of architecture at the University of Hartford, named Blount, who drew up a set of plans for a building that was rejected by the Hartford building inspection department because Blount was not a registered architect.

The South Carolina "architectural designer" was brought into court by the state, which sought an injunction forbidding him to practice architecture. The complaint listed a rather impressive group of buildings for which he had drawn plans and specifications and superintended the subsequent construction. For his services he was paid either a percentage of the cost of construction or a flat fee. He had always referred to himself as an "architectural designer" and it was generally understood that he was not a licensed architect.

The section of the South Carolina Code in question reads: "56-63. Chapter inapplicable to contractors, etc. Nothing in this chapter shall be construed to apply to contractors, builders, mechanics or private individuals making plans and erecting buildings, so long as they do not hold themselves out to the public as architects."

The trial judge enjoined the defendant from holding himself out as an architect (something that he claimed he never did), but allowed him to continue in business much as before. The court's order read:

(2) That the defendant is enjoined from practicing architecture or from holding himself out to be an architect unless he is properly licensed by the State Board of Architectural Examiners.

(3) That the defendant may continue carrying on his business as he has in the past, that is, drawing plans and erecting buildings, so long as he does not hold himself out to be an architect.

The state, dissatisfied with the second portion of the ruling, took an appeal to the Supreme Court of South Carolina.

That court agreed with the state's contention and reversed that part of the lower court's order permitting the defendant to continue his activities. The supreme court said in part:

> We do not think, however, that respondent is engaged only in making plans and erecting buildings within the usual and ordinary meaning of those words. Instead, he is engaged in making plans, writing up detailed specifications, and the supervision of the erection of buildings for the owners by contractors, rather than engaged simply in making plans and erecting buildings. Of course, supervision is, in a sense, a part, or at least contributes to the proper construction or erection of a building, but the contractors were the ones who actually erected the buildings. Thus we think that within the usual and ordinary meaning of the statutory language, respondent's activities did not come within the terms of the exemption. . . .
>
> In our view, the respondent was clearly practicing the profession of architecture without being licensed as such, in violation of law, and the lower court correctly enjoined him from practicing architecture or from holding himself out to be an architect unless properly licensed. The holding of the lower court, however, "that the defendant may continue carrying on his business as he has in the past, that is, drawing plans and erecting buildings so long as he does not hold himself out to be an architect," was, we think, for the reasons hereinabove set forth, erroneous, and to that extent, the decision of the lower court is reversed. To allow respondent to continue carrying on his business as in the past, was, in effect, an authorization to violate the injunction already granted in the previous paragraph of the order. (*State vs. Montgomery*, 136 S.E.2d 778.)

When the Connecticut carpenter–student's plans were not accepted because he was not an architect, and in addition, the plumbing, heating and electrical system details were not complete, he sought the assistance of a firm of structural engineers, who went to work on the necessary plans and specifications. Before their work was finished, the owners of the building abandoned their plans because they could not obtain the necessary financing, and the structural engineering firm brought an action against the carpenter–student asking payment for services rendered. He in turn brought in the corporation which owned the building as a co–defendant.

The trial court ruled that the plaintiffs could collect from the building owners for their unfinished work on the ground that the carpenter–student was their agent. An appeal was taken to the Appellate Division of the Circuit Court of Connecticut.

That court affirmed the ruling of the lower court in favor of the plaintiffs. It held that even though the carpenter-student could not practice architecture in his own right (it commented that if he had done so he would have subjected himself to criminal prosecution) he could, and did, act as an authorized agent for the owners of the building when he employed the plaintiffs. On this point the court said in part:

> Presumably, the defendant, as owner, instead of employing an architect could deal individually with those whose special professional skills were needed in order to comply with the statutes and ordinances pertaining to the erection of the proposed building. This task, the court has found, it delegated to Blount.
>
> At the very outset when Blount's sketches were rejected by the building inspection department as not meeting the legal requirements and he made known that fact to the defendant, the latter had the choice of discontinuing its operations through Blount and engaging the services of a registered architect; or of selecting qualified individuals to perform the necessary architectural and engineering services; or of having Blount make the necessary arrangements for such professional services as the defendant's agent. It adopted the latter course and, after having received the benefit of the work of the plaintiffs and those engaged by them, it seeks to disavow the agency.
>
> This it cannot do. "It is stated to be a well-established rule, that where a man stands by, knowingly, and suffers another person to do acts in his name, without any opposition, or objection, he is presumed to have given an authority to do those acts." (*Lapuk vs. Blount,* 198 A.2d 233.)

6. The Matter of Licensing

The public interest in the competent performance of engineering work has led to statutes requiring that engineers be licensed. The cases in this section consider the effects of such statutes and also contain some valuable definitions of *engineer* and of the scope of the work an engineer is permitted to perform in practicing his profession.

Are State Engineering Examination Boards Legal?

The power of Nevada's State Board of Registered Professional Engineers to conduct examinations and certify those who qualify as professional engineers was upheld by the state's supreme court.

The statutes empowering the board to so act were challenged by a man who, from 1953 to 1965, had been listed on the roster as a Professional Engineer–Land Surveyor. Then the board restricted his listing to Land Surveyor. He won a victory in the trial court, which declared the challenged statutes void and unconstitutional and enjoined the board from publishing a roster classifying those listed into the various branches of engineering. The board appealed the judgment to the Nevada Supreme Court.

That court stated the facts as follows:

> Upon a motion by the respondent Jack K. Leavitt, the trial court granted summary judgment and declared that NRS 625.180 through 625.210 inclusive, and NRS 625.520, were void and unconstitutional on the grounds that they are an illegal delegation of legislative authority.
>
> The lower court further ordered that the appellants be forever restrained and enjoined from classifying the respondent or any others similarly situated into any branch or branches of engineering, and prohibited the printing, publishing and distribution of a roster of profes-

sional engineers which classified members into branches of engineering. The order for summary judgment and judgment also required appellants to issue [to the] respondent and others similarly situated a registration card showing the respondent to be entitled to practice the profession of professional engineering without limitations as to any class or branch of the profession.

The respondent was graduated from Heald Engineering College in September of 1949. On October 5, 1951, he took and passed the Nevada engineer-in-training test. On July 7, 1953, he took the Nevada land surveyor examination. He passed only Part "A" of that test. On October 19, 1953, he took and passed Part "B" of the land surveyor examination. On November 16, 1953, he was approved as a registered land surveyor and issued a certificate. The respondent's title on this first certificate was phrased "Professional Engineer–Land Surveyor."

The court then noted that "In 1961 the legislature amended the statute governing the licensing of land surveyors so as to expressly exclude land surveying from the profession of engineering. Prior to this amendment, land surveyors had been a lesser category of 'engineers' and were excluded from that status only by implication."

(The amended statute read as follows: "The practice of land surveying does not include the design, either in whole or in part, of any structure of fixed works embraced in the practice of professional engineering. . . .")

The court's opinion continued:

The record shows that the respondent continued to be certified as a "Professional Engineer–Land Surveyor" until 1965. In 1966 he was issued a certificate which bore only the title "Land Surveyor."

On April 14, 1967, the respondent filed suit against the appellants requesting (1) that the board be enjoined from classifying him as anything other than a professional engineer; (2) that the board be enjoined from publishing a roster of professional engineers with listings according to the different branches of engineering; (3) that NRS 625.180-625.210 and NRS 625.520 be declared unconstitutional; and (4) that the board be required to "return" him to the status of professional engineer without limitation as to field of practice.

The appellants answered the respondent's complaint, and on June 16, 1967, the respondent moved for a judgment on the pleadings, or in the alternative, a summary judgment.

On July 28, 1967, the lower court granted Leavitt's motion for summary judgment. This appeal is taken from that order and judgment. As assignments of error the appellants claim that the lower court erred when it:

(1) Declared NRS 625.180 through 625.210 and NRS 625.520 unconstitutional.

(2) Enjoined the appellants from printing, publishing and distributing a roster of professional engineers that lists any classification into branches of engineering.

(3) Directed the appellants to issue the respondent and all others similarly situated, a registration showing their entitlement to practice the profession of professional engineer without limitation

(4) Entered its order for summary judgment.

The court first considered the question of the constitutionality of the challenged statutes:

> The summary judgment ordered by the trial court declared five separate statutory sections unconstitutional—NRS 625.180 to 625.210 inclusive, and NRS 625.520.
>
> NRS 625.180 lists, in detail, the qualifications necessary for admission to the ranks of professional engineering. Although the phrase "satisfactory to the board" is used, the content of the statute is clear and unambiguous. NRS 625.190 requires that the board hold examinations twice a year and determine, from the results, in which branch of engineering the applicant is qualified. NRS 625.200 delineates the scope of the examinations, the manner in which they are to be given, and what shall constitute a passing grade. NRS 625.210 simply directs that the board award certificates to all those meeting the requirements of the chapter. NRS 625.520 provides the penalties to be inflicted upon those who practice engineering unlawfully.
>
> The respondent argues that the discretion vested in the board, found in phrases such as "satisfactory to the board," "approved by the board" and "as determined by the board," is so unfettered as to render the statutes unconstitutional as an undue delegation of legislative authority.

The supreme court ruled that the delegation of authority to the board by the legislature was valid and that the challenged statutes were constitutional. In so doing, it relied principally upon the decision of the United States Supreme Court in *Field vs. Clark,* 143 U.S. 649, 12 S.Ct. 495, 36 L.Ed.294 (1892), in which the court said that "the true distinction . . . is between the delegation of power to make the law, which necessarily involves a discretion as to its execution, to be exercised in pursuance of law. The first cannot be done; to the latter no valid objection can be made."

The Nevada court also quoted from *Douglas vs. Noble,* 261 U.S. 165, 43 S.Ct. 403, 67 L.Ed. 590 (1923), in which United States Supreme Court

Justice Brandeis, in a case dealing with the qualifications of a dentist, said:

> The statute provides that the examination shall be before a board of practicing dentists, that the applicant must be a graduate of a reputable dental school, and that he must be of good moral character. Thus, the general standard of fitness and the character and scope of the examination are clearly indicated. Whether the applicant possesses the qualifications inherent in that standard is a question of fact. . . . The decision of that fact involves ordinarily two subsidiary questions of fact. The first, what the knowledge and skill is which fits one to practice the profession; the second, whether the applicant possesses that knowledge and skill. The latter finding is necessarily an individual one. The former is ordinarily one of general application. Hence it can be embodied in rules. The legislature itself may make this finding of the facts of general application, and by embodying it in the statute make it law. When it does so, the function of the examining board is limited to determining whether the applicant complies with the requirement so declared. But the Legislature need not make this general finding. To determine the subjects of which one must have knowledge in order to be fit to practice dentistry, the extent of knowledge in each subject; the degree of skill requisite; and the procedure to be followed in conducting the examination—these are matters appropriately committed to an administrative board. . . . And a legislature may, consistently with the federal Constitution, delegate to such board the function of determining these things, as well as the function of determining whether the applicant complies with the detailed standards of fitness.

The Nevada court continued, "Upon a thorough reading of the statutes under question, it becomes apparent that they enunciated the public policy as set forth by the legislature and established the basic standards to be required of engineers, vesting only a modicum of discretion in the board to carry out the policy and meet those standards."

After rejecting the reasoning of a Colorado case cited by the respondent, Leavitt, the court said:

> The power to examine applicants, formulate the questions to be asked, grade the examinations and determine the degree of skill required for a specific profession has been consistently held to be an administrative rather than a legislative function. [Citations]
>
> We find that the questioned statutes express a valid delegation of authority from the legislature to the state board of registered professional engineers. The trial court was in error when it declared NRS 625.180 through 625.210 and NRS 625.520, illegal, invalid, void and unconstitutional. . . .

We find that a roster may be printed, published and distributed in accordance with NRS 625.170. However, only the names of the registered professional engineers, land surveyors and engineers-in-training may be listed. Unless specific legislative authority is granted, the names of license holders may not be listed in branch classifications.

The respondent contends that he was originally licensed as a professional engineer and that the appellants, by issuing him a license as a land surveyor, are attempting to revoke his status without a full and proper hearing. The copy of the respondent's original certificate, which he attached as an exhibit to his complaint, belies this contention. His professional engineering status is qualified by the words "land surveyor."

The respondent is a licensed land surveyor and asserting that he is a professional engineer will not change his status.

The court then turned to the matter of the summary judgment granted by the trial court and, after discussing the general principles governing such judgments, said:

In the present case there was ample evidence presented by the appellants to compel the denial of respondent's motion for summary judgment.

The record contains certain documents showing that Leavitt had never taken the required second half of the professional engineers' examination. The most significant of these documents is a letter dated February 1, 1967, from Edward I. Pine, chairman of the Board of Registered Professional Engineers to the respondent, advising him that he had not taken the required "part two" of the civil engineer examination. In that letter the board states it would waive the 8 year statute of limitation on the engineer-in-training test (the first half of the engineering test) that Leavitt had taken and passed on October 5, 1951, and that he should present himself for the second half of the engineering test in Las Vegas on June 10, 1967.

Furthermore, there was, before the court, all of the correspondence between Leavitt and the board from 1951-1967, in which it is clearly established that Leavitt was always recognized only as a land surveyor, and that he never had taken the examination of all professional engineers. . . .

The question of the respondent's professional status was at issue. It was not only a material fact but the heart of the litigation, and the appellants had a right to present their evidence and be heard on this important question.

The judgment of the lower court is reversed and the case remanded for further proceedings consistent with this opinion. (*Pine vs. Leavitt*, 445 P.2d. 942.)

If There Is Such a Person as an Engineer, Can He Claim a Mechanic's Lien?

Presented in this section are two cases involving engineers. In the first case, the Supreme Court of Iowa appears to have made a valiant attempt to define the word *engineer,* but it reached a none-too-satisfactory conclusion. In the second, the Court of Appeals of Maryland, that state's highest judicial tribunal, decided that an engineer who prepares plans and specifications for a job is entitled to a mechanic's lien even though he performs no other service.

The plaintiff in the Iowa case was the State Board of Engineering Examiners, which sought an injunction to enjoin the defendant corporation from calling itself an engineering company. The district court denied the injunction and so the board appealed to the Supreme Court of Iowa. That court stated the facts as follows:

> From 1949 to 1958 William R. Clark and John J. Moffet, together with one other person, carried on a partnership business under the name of Electronic Engineering Company. In 1958 the business was incorporated under the same name. At the time of trial the business had been operated without interruption under this name for approximately 17 years, both as a partnership and as a corporation. The business of the defendant is principally that of installing and servicing two-way radio and mobile telephones. It also services and repairs home laundries, refrigerators, television sets, and air conditioning equipment. Defendant has never rendered professional engineering services, nor has it advertised that it was qualified to do so. Nor, in 17 years, has anyone requested defendant to render professional engineering services.

In describing the board's position, the court said,

> Plaintiff, however, contends what defendant does is unimportant and how it advertises is immaterial. The sole issue raised by the petition is whether the use of the corporate name Electronic Engineering Company, violates section 114.24 by *implying* defendant offers professional engineering services or by designating it as a professional engineering company.
>
> As pointed out in *State vs. Durham,* Del. Super., 191 A.2d 646 at 648, all 50 states, as well as the District of Columbia, the Canal Zone, and Puerto Rico, now have laws for licensing and registering professional engineers. Most if not all of these statutes have provisions governing the use of the words "professional engineer." Some specifically prohibit the use of the term at all in the title. [Citation] Other statutes, like ours, forbid the use of any name which designates a person as a professional engineer or implies he is one.

The latter type affords no fixed standard by which to decide if an injunction should issue. Whether a particular name raises the implication which the law forbids is a question of fact to be determined in each case.

The supreme court reiterated the trial court's position that the corporate name did not violate the statute. The lower court then dismissed the petition seeking to enjoin its use.

The supreme court said:

> Our review is *de novo* although we give consideration to the findings of the trial court. Rule 344 (f). R.C.P.
>
> Section 114.2, Code, 1966, defines professional engineer as a person who, by reason of his knowledge of mathematics, physical science, and the principles of engineering, acquired by professional education and/or practical experience, is qualified to engage in engineering practice.
>
> That section defines the practice of professional engineering to mean any professional service, such as consultation, investigation, evaluation, planning, designing, or responsible supervision of construction in connection with structures, buildings, equipment, processes, works, or projects, wherein the public welfare or the safeguarding of lives, health or property is or may be concerned or involved, when such professional service requires the application of engineering principles and data.
>
> Neither of the individual defendants is a professional engineer. The corporate defendant does not render, nor does it hold itself out as rendering, professional engineering services as defined in the statute.
>
> Our problem is to determine if the corporate name implies that it does.

In the court's research of the definition of *engineer,* it discovered that the word "has now lost much of its original professional significance."

The court continued:

> The evidence shows that the yellow pages of the Des Moines telephone directory list engineering in categories such as architectural, chemical, civil, consulting, electrical, heating, mechanical, sanitary, structural and traffic. There is no separate listing for electronic engineers. To these may be added other definitions from *Webster's International Dictionary, 2nd Edition,* which includes within the term one who runs an engine, one who manages a project by artful contrivance; and, in military parlance, one engaged as an artilleryman or gunner. This same dictionary notes there are 41 special classifications under the definition of engineering and even then indicates there are still others not listed.
>
> We mention this merely to illustrate that "engineer" no longer

necessarily connotes professional competence or skill. Apparently the legislature recognized this by limiting the application of Chapter 114 to *professional* engineers, which Section 114.2 defines to exclude many of those who now use that term to describe their work or occupation. It must be conceded that "engineer" and "professional engineer" are not synonymous. The use of one does not necessarily imply the other.

The sole issue here is whether defendant's corporate name implies defendant renders professional engineering services, not whether it renders *any* engineering services. In considering that question, it is important to take into account all the circumstances surrounding the use of such name. This is particularly true since our finding is one of fact and prior decisions from other jurisdictions are of no help to us.

Under the evidence here we agree with the trial court [that] there is no violation of Section 114.24. In reaching this conclusion we cannot completely disregard, as plaintiff would have us do, the testimony indicating that defendant has used this name for 17 years without receiving a single request to render professional engineering services. We consider this quite important in deciding the name does not imply what plaintiff says it does. The absence, over a long period of time, of a single request to render the type of service prohibited by statute overwhelmingly suggests the name raises no implication of [being] in violation of Section 114.24. It is the public, not the plaintiff, which the law seeks to protect. The evidence makes it clear what plaintiff argues here has occurred to no one except plaintiff.

We do not say that the use of a name such as defendant uses could under no circumstances constitute a violation of the section in question. As previously determined, this is a fact question to be determined in each case. We merely hold plaintiff has failed to show by a preponderance of evidence a violation of section 114.24, Code, 1966. (*Iowa State Board of Engineering Examiners vs. Electronic Engineering Co.*, 154 N.W.2d 737.)

In the Maryland case, an engineer who had prepared mechanical plans and specifications for a fifteen–story office building brought an action against the owners of the building to foreclose a lien for his services. The judge in the trial court ruled in favor of the owners, and the engineer appealed to the Maryland Court of Appeals. That court's opinion follows:

Judge Raine [the trial court judge] granted summary judgment for the owner of a 15–story office building in Towson in an action to foreclose a mechanic's lien duly filed by an engineer, Liebergott, who at the instance of the architect had prepared structural, electrical and mechanical plans and specifications for the building, some of which were used in the construction. Several months after work had started, Liebergott's services were dispensed with and another engineer's plans used thereafter.

The appellate court noted that the trial judge determined that Liebergott was a design engineer employed by the architect and that the contract filed as an exhibit with the motion provided "the engineer's responsibility for services during the construction phase of the work has been removed from this agreement. . . ." As further evidence, the trial record showed that in a letter dated Oct. 20, 1965, to Charles B. Wheeler, a Baltimore County official, Liebergott said "my services, however, for field supervision and inspection were not and are not a part of the contract." In that letter and in a subsequent one (Nov. 3, 1965) to the general contractors, the trial court's opinion continued, Liebergott specifically denied responsibility for any field supervision and inspection. The affidavits of the owner and of the general contractor and architect stated that Liebergott did not perform any supervision during construction of the building.

Despite the letters and contract, the record shows that Liebergott made affidavit that he performed supervisory and inspection services as part of his contractual agreement. He filed an exhibit showing that he made ten visits to the site that involved "inspection of progress of work." The trial court concluded that, since the building is fifteen stories high, "it is clear that the inspection trips by Liebergott could not possibly amount to superintendence or supervision of construction."

The appellate court noted that the trial court based its decision on *Caton Ridge, Inc. vs. Bonnett,* 245 Md. 268, 225 A.2d 853. The appellate court quoted this portion of the lower court's opinion: "Reading and rereading that case convinces the court that an architect or design engineer is not entitled to a mechanic's lien unless he also supervises the construction of the building. The Court of Appeals adopted a general rule that an architect who furnishes plans *and* supervises the construction is entitled to a lien [Emphasis added] . . . . By any reasonable definition of the terms, supervision or superintendence, Liebergott did neither, and therefore is not entitled to a lien.

The appellate court, however, stated:

The case was not one for summary judgment. Under *Caton Ridge, Inc. vs. Bonnett* . . . a design engineer who prepares plans used in a building and supervises the use of his plans would be entitled to the protection of the mechanic's lien statute. Liebergott's contract with the architect did provide that his responsibility for services during the construction phase of the work has been removed from the agreement, but the contract went on to say:

"The engineer agrees to perform the following service without additional compensation:

"He shall check and approve shop drawings and make himself

available for consultation on any problem arising as a result of an error or omission generated by his office and make the necessary changes to the drawings to effect the correction.

"Any additional services shall be taken care of as described in Article 24 above."

The court's opinion then quotes at considerable length portions of affidavits made by Liebergott regarding the numerous visits that he made to the site at the request of the owner and his methods of keeping an account of such visits. The court then continued:

> It is apparent that a genuine dispute as to material facts was created by Liebergott's sworn statement that he did in fact supervise the use in the building of the plans he prepared. It was error to hold as a matter of law that "the 10 inspection trips could not possibly amount to superintendence or supervision of construction." Liebergott was dismissed at a relatively early stage of construction and could supervise or inspect only for a few months, and during those months his visits to the site were relatively frequent. In *Caton Ridge* the architect was on the site at first weekly, then every two weeks and later once a month. In the case of *Chesnow vs. Gorelick,* 246 Mich. 571, 225 N.W. 4, 6 (1929), relied on in *Caton Ridge,* the court rejected the owner's contention that the architect could not have a lien because his inspections took but a small part of his time and he did not actually superintend construction. The court said the contract permitted the architect to determine the number of visits to the site properly to work his plans into the building.

The appellate court concluded:

> Since there was an issue of fact raised by the pleadings and affidavits, summary judgment could not rightly be granted. . . .
>
> Since the decision on remand may turn entirely on whether Liebergott comes factually under *Caton Ridge,* we need not and do not decide whether a design engineer who under contract prepares plans for a building, which are used in the construction of the building, but who does not supervise or superintend construction is entitled to a mechanic's lien for unpaid services.

Although declining to rule on the above point, the court concluded with an informative discussion of the law on the subject in other jurisdictions:

> It is appropriate, nevertheless, to say that in our view Judge Raine

read too much into *Caton Ridge* when he found it to hold that an architect who prepared plans but did not supervise was not within the coverage of the mechanic's lien statute. That question was neither decisive nor decided in *Caton Ridge* since there the architect did both, and all we needed to decide and all we did decide was whether an architect who did both was covered.

It should be noted that an architect or engineer who under a contract prepares plans for a building which are used in that building is within the literal coverage of Code (1968 Repl. Vol.), Art. 63, Sec. 1, which gives a lien on a building "for the payment of all debts contracted for work done for or about the same. . . ." Preparation of plans used in the erection of a building literally is work done for the building. . . .

The cases throughout the country seem to be about equally divided where there is preparation of plans which are used but no supervision or superintendence. See the cases collected in the annotation in 60 A.L.R. 1257, 1267 (1927). Some of the older cases referred to allow a lien in such situations on the ground that preparing plans was not manual labor and the lien statute was intended to protect only mechanics and artisans who furnished physical labor. That ground of construction of the lien statute would not seem tenable in Maryland. . . .

An excellent analysis of the cases both ways is to be found in *Gaastra, Gladding & Johnson vs. Bishop's Lodge Co.*, 35 N.M. 396, 229 P. 347 (1931), which concluded that a non–supervising architect was entitled to a lien where under contract he prepared plans that were used in the construction of a building.

Judgment reversed and the case remanded for further proceedings.

Thus Liebergott, the engineer, was given another chance to prove that the work he did entitled him to a mechanic's lien under the Maryland statute and the interpretation thereof in this case. (*Morris J. Liebergott & Associates vs. Investment Building Corp.*, 241 A.2d 138.)

When Is Engineering Not Engineering?

The Wisconsin Registration Board of Architects and Professional Engineers sought an injunction to prevent a corporation known as T.V. Engineers, Inc. of Kenosha from using the word *engineers* in its name and in its advertising. The circuit court granted the plea for an injunction and the defendant corporation appealed the ruling to the Supreme Court of Wisconsin.

In a four-page decision in which the Wisconsin statutes governing the registration of professional engineers were quoted at length, the court said in part:

The word "engineer" or "engineering" used in a business title in its descriptive context, could tend to convey the impression that it is engaged in the practice (or offers to practice) the profession of professional engineering. "Engineer" or "engineering" when used with such descriptive words as civil, electrical, mechanical, industrial, structural and others could well tend to convey an impression of professional engineering service. Whether the use of "engineer" or "engineering" in a business title tends to convey the impression of practicing or offering to practice professional engineering must then be determined as a matter of fact by the circumstances of the case under consideration.

In this case before us the evidentiary facts, in the main, are not in dispute. It is the conclusive ultimate fact to be drawn from the evidentiary facts that is disputed.

The defendant's business is primarily the sale, installation and service of television sets and other electrical appliances. Their overwhelming emphasis of the rather voluminous advertising is in the sale of appliances.

In none of the advertising is professional engineering advertised as such. Nor has the defendant corporation practiced any professional engineering unless repairing can be considered as such.

Both Mr. Young, a registered professional engineer, and Mr. Hurc, the secretary of the registration board, testified that improper repair of certain types of television sets can result in a condition that can be dangerous if not fatal to a member of the public. It also appears that many commonly used electrical appliances can be dangerous if not properly repaired. It is basic that such repair work requires some knowledge of either mathematics or physics, or both.

Mr. Hurc testified that television repair men need not be registered with the board; in fact, the board does not have a designation for television in its classification of professional engineers. Mr. Hurc was indefinite as to what category would encompass television repair. He stated that television might be within one of the five sub-fields of electrical engineering and that two sub-fields, communications and industrial electronics, would fit the television industry. He stated that examinations in neither of these sub-fields took television repair into consideration.

The current telephone directories of both Kenosha and Madison were introduced in evidence. "T.V. Engineers" and "T.V. Engineers of Kenosha, Inc." appear at least four times in the classified section—the yellow pages of the Kenosha directory. It is significant that defendant's ad and listings appear in the section classified as "Television" and not in the section classified as "Engineers" or "Engineering."

The sub-classifications for engineers in the Kenosha directory are architectural, consulting, designing, management and structural. In Madison the same sub-classifications appear, and in addition thereto civil, electrical, electronic, foundation, mechanical, metallurgical and sanitary. Mr. Young is listed under the "Engineers" classification as "Electronic" and under the "Television" classification as "Television Repairing." His yellow page ads also state he is licensed and bonded.

We are of the opinion, under the facts as they appear in the record, that the use of the word "engineers" in defendant's corporate name "T.V. Engineers" or "T.V. Engineers of Kenosha, Inc." did not tend to convey the impression that it is engaged in the practice of the profession of professional engineering, nor did it advertise to furnish professional engineering service. The finding of the trial court to the contrary is, in our belief, against the great weight and clear preponderance of the evidence.

Judgment reversed and cause remanded with directions to dismiss the complaint. (*State vs. T. V. Engineers of Kenosha, Inc.* 141 N.W. 2nd 235.)

Who Pays For an Abandoned Project?

When an almost $2 million public construction project is abandoned before actual work on the site is begun, what happens to the engineer who has prepared the plans and specifications for the project? That was an issue tried before the Appellate Court of Illinois, First District, Fourth Div. The court rendered a judgment in favor of the engineer.

The plaintiff in the action was Walter D. Wilson, a Chicago civil and mechanical engineer, and the defendant was the Village of Forest View, Ill. The appellate court described the situation as follows:

The Village of Forest View appeals from a judgment entered in the Circuit Court of Cook County against it and in favor of the plaintiff in the sum of $67,292.40. On the original hearing the defendant raised only two questions: (1) "Was there a valid contract between plaintiff and defendant?" (and) "(2) Did defendant abandon the Water Works Improvements and Sewer System for which the plaintiff had drawn plans which were given to and used by the defendant?"

From the record it appears that on November 10, 1959, the plaintiff filed a complaint in two counts against the Village of Forest View, a municipal corporation. The Village filed an answer, and a reply to the answer was filed by plaintiff. Count 1 was based on an oral contract allegedly entered into between the plaintiff and the Village in "the early part of November 1955." This allegation was denied by the defendant.

The second count was based on the allegation of plaintiff that a subsequent contract was entered into on April 22, 1957, reducing to writing the verbal agreement which allegedly had been entered into between plaintiff and the Village. The defendant admits that there was such a written agreement but denies that the agreement was legally entered into, and further denies that the written instrument was a reduction to writing of the previous oral agreement between the plaintiff and the defendant.

From an inspection of the pleadings it appears it was admitted by the Village that the plaintiff, Walter D. Wilson, a resident of Chicago, Ill., was a duly licensed civil engineer engaged in the practice of that profession, specializing as a consultant and mechanical engineer, having been so engaged for the past 25 years; that the Village, in accordance with the law, proposed certain local improvements to be made in said Village, which improvements involved the construction of certain water mains, reservoirs, and a pumping station and supply line, hereinafter referred to as the Water Works Improvements, and for the construction of the combined sanitary and storm sewer system, which will hereafter be referred to as the Sewer System; that in section 5 of his complaint the plaintiff alleges that "pursuant to [his] employment" he prepared all the necessary plans, specifications, etc. In its answer the Village states, "Defendant admits the allegations contained in Paragraph 5 of Count 1 of Plaintiff's complaint."

The total estimate of costs for the Water Works Improvements aggregated $942,193, which included the sum of $65,953, for the costs of engineering services for planning and supervision of construction, or a total of $819,708, excluding costs of engineering services, and making, levying and collecting the special assessments. The total estimate of costs for the Sewer System aggregated the sum of $997,228, which included the sum of $74,792 for the costs of engineering services for planning and supervision of construction, or a total of $862,602, excluding the costs of engineering services for planning and the costs of making, levying and collecting special assessments.

Ordinances for the Water Works Improvements and Sewer System were presented to the President and Board of Trustees of the defendant Village, together with the recommendations of the Board of Local Improvement for such improvements. On December 19, 1956, the said ordinances were adopted and approved by the President and Board of Trustees of the Village. On January 11, 1957, plans, specifications and estimates of costs prepared by the plaintiff and approved by the Village were filed in the County Court of Cook County, pursuant to Local Improvements Act (Ill. Rev. Stat. 1955, Ch. 24, Art. 84), and identified as Special Assessments Nos. 5 and 6 of the Village of Forest View, for the construction of the Water Works Improvements and Sewer System.

At this point it would seem as though the plaintiff engineer was reasonably assured of collecting his fees for his services under the oral and written contracts. Every provision of the statutes seemed to have been complied with.

But there was trouble ahead. At the time the contract was entered into, it was contemplated that the work would be paid for out of special assessments for that purpose; but various lawsuits developed, including one that sought to disconnect certain areas that were involved in the Water Works Improvements and Sewer System from the Village. Further complications were caused by an election in 1957 when the president of the Village and three members of the Board of Trustees were defeated and control of the Village was transferred to a newly elected majority. One result was the abandonment of the construction plans.

A provision of the Illinois statutes was relied upon by the Village authorities to avoid payment of the plaintiff's claim. They contended that they were forbidden to pay the claim out of the general fund by this provision which provided that such fees be paid out of the proceeds of the special assessment that was contemplated when the contract was made. A long list of Illinois cases was cited in support of this view.

But the court carefully distinguished these cases from the one before it and said:

> We conclude that since in the instant case, under the provisions of Chapter 24, Sec. 8-1-7 (in force at the time the contract in question was entered into), the Village could have legally executed a contract for the improvement which was payable out of a special fund, there was no necessity for any appropriation being made out of the general fund at the time of the inception of the contract. Indeed, such an appropriation ordinance would have been inconsistent with the terms of the contract.
>
> When the Village, after utilizing the product of the plaintiff's services, thereafter abandoned the project, it prevented creation of the fund out of which the plaintiff was to have been paid, but it could not unilaterally impair the validity of the contract. It therefore remained liable to meet the obligations thereunder out of the general fund.

In further support of its ruling that the contract was valid, the court recited the following facts in regard to the action taken by the Village officials:

> The Village urges that since there was no record of a meeting of the Board of Trustees of the Village in November, 1955, the trial court was not justified in finding that an oral contract was entered into at that time between the plaintiff and the Village.
>
> The trial court had a right to consider the paucity of records of proceedings of the Village and to take judicial notice of the subsequent

court proceedings, both civil and criminal, brought against certain officials of the Village. The plaintiff and several persons who were members of the Village Board in 1957 testified that the oral contract alleged in the complaint had been entered into at a meeting of the Village Board of Trustees in November, 1955. This was contradicted by other than members of the Board of Trustees, and while no record was introduced in evidence showing that the Board of Trustees passed a motion or discussed the question, it was admitted that in many instances there were no records in existence showing the proceedings of the Village Board.

In our opinion, the court was fully justified in finding that an oral agreement had been entered into between the Village and the plaintiff. There is no dispute that the Village Board purported to ratify an oral contract on April 22, 1957, and it is admitted that the Village used the plans, specifications and estimates of cost prepared by the plaintiff in its special assessment proceedings in the County Court of Cook County.

On rehearing, the defendant Village quarreled with the method by which the trial court had calculated the amount of damages due, but the method was upheld by the appellate court, which said:

> Furthermore, in the instant case, the plaintiff testified that the usual and customary charges for the preparation of plans and specifications and estimates of cost in the County of Cook in the year 1955 were 4% of the estimated cost of construction for the preparation of the preliminary plans, final plans and specifications.
>
> The plaintiff further testified that he had received no payment from the Village, but his out-of-pocket expenses were about $20,000. At the time plaintiff was questioned by his counsel about the customary charge an objection was interposed by counsel for the Village, who subsequently withdrew his objection and stated that they would not object to the testimony as to 4%. Hence, there was proof in the record both of the amount which reasonably could have been charged, and this evidence was sufficient to support the amount of the trial court's judgment.

The judgment of the trial court in favor of the plaintiff in the amount of $67,292.40 was affirmed. (*Wilson vs. Village of Forest View*, 217 N.E. 2d. 398)

Should an Engineer's Fee Hinge on Licensing?

"Is a professional engineer's claim for consulting engineering services, rendered in New Jersey, to a New Jersey licensed architect, pursuant to contract, unenforceable as a matter of law by reason of the fact that the claimant was not licensed by the State of New Jersey as a professional engineer . . .?"

That interesting and important question is the beginning of an opinion handed down by the United States Court of Appeals, Third Circuit. (The federal courts had jurisdiction over the matter because the parties in this suit were from different states.)

The plaintiff in the case was Milton Costello, a professional engineer specializing in swimming pool design and the supervision of the pools' construction. He is licensed in New York—where he resides and has his place of business—and in the states of Maryland, Illinois and New Mexico; but he is not licensed in New Jersey. When a district court ruled against his collecting his claim, Costello appealed.

The defendant was architect Emil A. Schmidlin, who resides and has his business in New Jersey. A third-party defendant, the Township of Maplewood, is a New Jersey municipality.

As stated in the appeals court opinion:

> On August 18, 1965, defendant entered into a written contract with the Township for the installation and construction of a municipal swimming pool complex. At the Township's request, he thereafter, in October, 1965, entered into a written contract with plaintiff for his performance of certain engineering and other services in connection with the swimming pool project. This contract provided as follows: $3,000 as a retainer; $6,000 upon completion of plans; $3,000 upon award of contract or 60 days after issuance of plans and specifications for bidding, whichever occurred earlier; and $3,000 progress payments, pro rata, during the course of construction.
>
> The Township, by a Supplemental Contract with defendant, agreed to reimburse him for one-half of plaintiff's $15,000 compensation. In accordance with that contract it reimbursed defendant for one-half of the $3,000 retainer fee he paid to plaintiff.

The court then recited the contractual relationship of the parties and what services were performed:

> The plaintiff-defendant contract was captioned "Consulting Engineering Services Agreement." It expressly stated that defendant had a contract with the Township for furnishing complete professional services with respect to the construction of the swimming pool project, and that plaintiff's relationship to defendant should ". . . at all times be that of an Associate Consultant. . . ." It described defendant as "Client" of plaintiff and stated that ". . . the Client shall, without limit, have final right of review and approval of all plans and specifications which shall be the essence of this contract."

The plans and specifications were completed and issued for bidding on Jan. 12, 1966. The following day, Costello billed the architect for the $6,000 that was payable at that time. When bids for the job were received, they were substantially higher than the Township's estimate and more than its available funds. The architect requested that Costello revise the plans and specifications to fit the available appropriation in contemplation of a re-invitation for bids. Costello, in collaboration with the architect's staff, prepared and submitted changes. During the preparation of these revisions, the Township invited the consulting engineer to attend a meeting of its governing body; at the meeting a different type of pool was considered and recommended by the Township's governing body and engineer. Costello disapproved the substitute but to no avail: it was approved and the bid of the manufacturer of the substituted pool was accepted by the Township.

"On March 12, 1966," the appeals court said, "plaintiff unsuccessfully billed defendant for the $3,000 payment 60 days after issuance . . . of the plans and specifications for bidding. Later, he billed defendant for the unpaid installments of $6,000 and $3,000 due on his contract, and an additional $8,000 for professional services not called for in the contract. Still later, he submitted an additional bill for the $3,000 final payment provided by his contract."

When no payments were forthcoming, Costello sued the architect for the $12,000 billed under his contract and the $8,000 extra services.

The court described the architect-defendant's position and the trial court's reasoning:

> Defendant, in his "Answer and Counterclaim" alleged that "plaintiff was in fact not able, competent, duly qualified and licensed to perform" the services specified by his contract and that he "did not in fact perform in accordance with all the material terms and conditions of his agreement." He also filed a two-count counterclaim against plaintiff; in Count I he sought recovery of the $3,000 retainer fee paid to plaintiff; and in Count II he claimed plaintiff owed him $12,000 for services rendered to plaintiff which he alleged were plaintiff's responsibility under their contract.

> Defendant also filed a third-party counterclaim against the Township alleging therein that if he were held liable to plaintiff for the $12,000 contractual balance, Township was liable to him for one-half of that amount under the Township-defendant contract.

> Defendant's evidence at the trial established the plaintiff's billing pursuant to his contract had not been paid only because Township had not prior thereto paid defendant, even though their payment had not been conditioned by the contract upon Township's prior payment to defendant.

In an oral judgment, the district court, after ruling that Costello was not barred by New Jersey law from prosecuting his action, awarded him a judgment of $9,000, which it found was due him under his contract, and an additional $350 in *quantum meruit,* for work done on revisions of plans.

In a subsequent written opinion, however, the district court ruled that the contract between the engineer and architect was "unenforceable here as contrary to the laws of New Jersey." The reason for this reversal was the fact that the engineer did not have a New Jersey professional engineer's license.

The district court held *seriatum* that the engineer could not maintain his action; the architect was entitled to recover the $3,000 retainer he had paid the engineer; and "with respect to the second count of defendant's counterclaim against plaintiff [for $12,000], there is an absence of evident supportive of any right to recovery thereunder."

The appeals court stated that the district court "(1) entered judgment, with costs, in favor of defendant and against plaintiff, and dismissed plaintiff's complaint 'with prejudice'; (2) entered judgment in favor of the third-party defendant Township against defendant and dismissed the latter's complaint 'with prejudice'; (3) entered judgment in favor of the defendant against plaintiff on the first count of defendant's counterclaim in the sum of $3,000 with interest; and (4) dismissed the second count of defendant's counterclaim against defendant."

The appeals court then took up the question of who a registered professional engineer is:

> On this appeal, plaintiff contends that the District Court erred in ruling that he is barred from any recovery for his services as consulting professional engineer because he was not licensed as a professional engineer in New Jersey, and asks that the District Court's oral judgment in his favor in the amount of $9,350, announced from the bench at the conclusion of the trial, should be reinstated.
>
> Defendant contends to the contrary and urges affirmance of the District Court's final Order of Judgment, earlier stated.
>
> What has been said brings us to the question as to whether plaintiff's action was unenforceable as a matter of law under the prevailing facts by reason of the circumstances that he was not licensed by the State of New Jersey as a professional engineer. . . .

After quoting at length from New Jersey statutes governing the licensing of professional engineers, the court proceeded to a discussion of two New Jersey cases on which the district court had relied in reaching its contradictory decisions. The first of these, *Weinrott,* decided in 1952, construed the statute as precluding the recovery of money for engineering services rendered by a person not duly licensed. The second, *Magasiny,* decided in 1966, ruled that under certain circumstances the unlicensed engineer could collect.

The court of appeals pointed out that neither of these cases involved a situation exactly like the case before it.

In regard to *Weinrott,* it said, "It is apparent from the foregoing statement that *Weinrott* is inapposite here on its facts in these respects: (1) there plaintiff sought compensation for services rendered as *principal* professional engineer to a defendant engaged in housing operations, while here plaintiff sought compensation for services rendered as a *consulting* professional engineer to a New Jersey licensed architect; and (2) there plaintiff was not licensed as a professional engineer in *any* state while plaintiff was licensed as a professional engineer in New York at the time he contracted with defendant."

In differentiating *Magasiny* from the case before it, the court quoted from the older case:

> "It may be that the transaction between the parties did not require plaintiffs to be licensed, even though their other activities did. In that case, should not the defense fail? There are numerous exceptions to such statutes. . . ."
>
> 6A Corbin, *supra,* says:
>
> ". . . the statute may be clearly for protection against fraud and incompetence; but in very many cases the statute breaker is neither fraudulent nor incompetent. He may have rendered excellent services or delivered goods of the highest quality, his noncompliance with the statute seems nearly harmless, and the real defrauder seems to be the defendant who is enriching himself at the plaintiff's expense. . . ."
>
> It must be remembered that in most cases the statute itself does not require these forfeitures. It fixes its own penalties, usually fine or imprisonment of minor character with a degree of discretion in the court. The added penalty of nonenforceability of bargains is a judicial creation. In most cases it is wise to employ it; but when it causes great and disproportionate hardship its application may be avoided. . . .
>
> Furthermore, as Corbin points out, *supra,* even when the activity is literally within the statute, the facts may present a situation so border line or unfair that the defense of illegality will not be permitted, or, at least the defendant will not be allowed to keep what he received without making fair compensation.

Having demonstrated that in the *Magasiny* decision New Jersey courts had adopted a more liberal point of view, the court of appeals next quoted from a Kentucky decision which stated that although the general rule is that a contract cannot be enforced by an unlicensed person, there are exceptions. The Kentucky court said, quoted from a federal court decision:

But that general rule does not have application in a case of this kind in which an unlicensed member of a profession or trade seeks to recover from a licensed member for services rendered or labor performed pursuant to a contract entered into by them.

This exception seems to have been recognized in Oklahoma, Washington, New York, Colorado and California. . . .

The statute involved, and similar ones, are designed to protect the public from being imposed upon by persons not qualified to render a professional service. The reason for the rule denying enforceability does not exist when persons engaged in the same business or profession are dealing at arms length with each other. In the case before us, appellant was in a position to know, and did know, the qualifications of appellee. No reliance was placed upon the existence of a license, as presumptively would be the case if appellee was dealing with the general public.

The court also discussed a decision in a New York case that said, "Illegal contracts are generally unenforceable. Where contracts which violate statutory provisions are merely *malum prohibitum,* the general rules do not always apply. If the statute does not provide expressly that its violation will deprive the parties of their right to sue on the contract, and the denial of relief is wholly out of proportion to the requirements of public or appropriate individual punishment, the right to recover will not be denied."

The court of appeals continued: "It must be noted at this juncture that the New Jersey professional engineers' licensing act . . . does not contain a provision barring enforcement of claims of nonlicensed engineers, and only provides for imposition of fines for its violation. Significantly, New Jersey's real estate brokers' licensing act was amended in 1953 to bar claims for compensation of nonlicensed brokers."

The outcome of the court's citing possible reasons why the plaintiff engineer should be compensated, despite the fact that he had no New Jersey professional engineer's license, was a reversal of the written ruling of the trial court. The appeals court awarded engineer Costello $9,350. In so doing it pointed out that plaintiff, as a professional engineer, and defendant, as an architect, are persons in the same business field who have contracted with knowledge of each other's respective professional qualifications, and that no reliance was placed by architect on the engineer's having a New Jersey license, since the engineer–architect contract provided only that Costello furnish consulting engineering services to the architect.

The court concluded:

Upon consideration of the factual elements and the factor that the New Jersey statute here involved does not contain a provision barring enforcement of claims of nonlicensed engineers, we are of the opinion that plaintiff's claim for consulting engineering services against defendant is not unenforceable by reason of the circumstance that he was not licensed to practice as a professional engineer in New Jersey and that the Supreme Court of New Jersey would so hold. (*Costello vs. Schmidlin*, 404 F.2d 87.)

7. The Engineer's Responsiblities

Some of the duties and responsibilities an engineer assumes, or inherits, when he accepts employment in some phase of a construction job are set forth in the cases in this section. Although ostensibly the representative of the owner, to whom he owes his principal allegiance, the engineer is a favorite target of all of those engaged in the building project.

Do the Engineer's Responsibilities Include Safety of Contractors' Equipment?

The extent to which an engineer may be held responsible for mishaps on a construction job was the issue in a case that the Supreme Court of Arkansas decided. An injured worker sought to recover damages from the consulting engineer, Forrest & Cotton Inc., and the resident engineer, Garver & Garver Inc., on a sewer project in Little Rock. Though a jury brought in a verdict in the workman's favor for $7,500, the court granted a motion for a judgment for the engineers. The worker, Jack L. Heslep, appealed to the supreme court.

In arguing for a reversal of the trial judgment, Heslep, who was an oiler, driver, and hookup man on a mobile crane, said the only issue was whether there was substantial evidence from which the jury could find that the engineers had a duty to require the contractor to provide safety devices on his mobile crane while it was operated close to a 13,000–volt power line. The supreme court found no such evidence and upheld the trial court's judgment.

This is the background for the case:

Heslep's employer, Paul N. Howard Co., had a general contract with the City of Little Rock to install a new underground sewer system. The system was to be installed according to plans and specifications prepared by an out-of-state engineering firm, Forrest & Cotton Inc. Garver & Garver Inc., a local firm, was employed to see that plans and specifications were complied with. One section of the sewer system was built of joints of pipe five feet long and three feet in diameter. These were lowered into a ditch and cemented into place. The joints of sewer pipe were shipped in from out of state and delivered by truck. The pipe joints were inspected for cracks and flaws when they arrived, and all rejected joints were returned to the manufacturer when the trucks went back for more pipe.

On the day Heslep was injured, a truck had delivered a load of pipe and by noon was ready to leave on the return trip. Two rejected joints of pipe were to be returned, and by the use of a front-end loader, they had been moved between a sidewalk and street curb, which was about 30 feet beneath an electric power line strung on light poles, also between the side-walk and curb. The truck was too high for the pipe to be placed on it by the loader, so the loader was used to move the rejected pipe joints from under the power line and out into the street, where they could be picked up by the mobile crane and put on the truck.

Eddy Faulkner, another employee of Paul N. Howard Co., operated the crane boom. Heslep was supposed to drive the crane from place to place and spot it as Faulkner directed; Heslep would then fasten the cable from the crane boom to the object and release the cable after the object had been moved.

On the day Heslep was hurt, the two employees' foreman, R. L. Welsh, had ordered them to hoist the rejected pipe onto the truck. The crane was moved into position about 12:30 P.M. The crane was equipped with either a 50-foot boom, as testified by Welsh, or a 70-foot boom, as testified by Faulkner, and the joints of pipe were lifted by a swing cable attached to the joints of pipe and also to the lift cable on the crane.

One joint of the rejected pipe had been moved into the street by the front-end loader when the loader operator went to lunch, leaving the other joint under the power line. When Heslep and Faulkner arrived with the crane, they loaded the joint of pipe from the street onto the truck. Because the truck would be ready to leave as soon as the other joint of pipe was loaded, Heslep and Faulkner did not wait for the loader operator to return from lunch. Instead, Heslep pulled or carried the swing cable, still attached to the lift cable on the crane boom, to the joint of pipe under the power line. As he was trying to fasten the sling cable to the pipe, the lift cable, or boom, came in contact with the power line, and Heslep was injured.

Some of the provisions in the contractual arrangements between the parties on which Heslep seemed to rely are as follows:

> In case it is necessary to change or move the property of any
> owner or of a public utility, such property shall not be moved or inter-

fered with until ordered to be done so by the engineer. The right is reserved to the owners of public utilities to enter upon the limits of the project for the purpose of making such changes or repairs of their property that may be made necessary by performance of this contract. . .

It is further agreed that if the work or any part thereof, or any material brought on the site of the work for use in the work or selected for the same, shall be deemed by the engineer as unsuitable or not in conformity with the specifications, the contractor shall, after receipt of written notice thereof from the engineer, forthwith remove such material and rebuild or otherwise remedy such work so that it shall be in full accordance with this contract. . . .

The contractor shall give personal attention to the faithful prosecution and completion of the contract and shall keep on the work, during its progress, a competent superintendent and any necessary assistants, all satisfactory to the engineer. The superintendent shall represent the contractor in his absence and all directions given to him shall be as binding as if given to the contractor. Important directions shall be confirmed in writing to the contractor. Other directions shall be so confirmed on written request in each case. . . .

The contractor shall take out and procure a policy or policies of workmen's compensation insurance with an insurance company licensed to do business in the State of Arkansas, which policy shall comply with the Workmen's Compensation Law of the State of Arkansas. The contractor shall at all times exercise reasonable precautions for the safety of employees and others on or near the work and shall comply with all applicable provisions of federal, state and municipal safety laws and building and construction codes. All machinery and equipment and other physical hazards shall be guarded in accordance with the "Manual of Accident Prevention in Construction" of the Associated General Contractors of America except where incompatible with federal, state or municipal laws or regulations. The contractor shall provide such machinery, guards, safe walkways, ladders, bridges, gangplanks and other safety devices as may be required by the engineer as requisite to the prevention of accident. . . .

Heslep also quoted from Arkansas statute Ann. Sec. 81-1406 (Supp. 1967) as follows:

The operation . . . of any . . . machinery or equipment . . . capable of vertical, lateral or swinging motion . . . by or near overhead high-voltage lines, shall be prohibited if at any time during such operation . . . it is possible to bring such equipment . . . within six feet of such overhead voltage lines, except where such high-voltage lines have been

effectively guarded against danger from accidental contact, by either:
(1) the erection of mechanical barriers to prevent physical contact with
high-voltage conductors: or (2) de-energizing the high-voltage conductors and grounding where necessary. Only in case of either of such exceptions may the six-foot clearance required be reduced.

Heslep contended that under the conditions there was a requirement
that, before the pipes were removed, the power line should have been de-
energized; that the contractor's agreement with the owner, which the engineers
prepared, provided that if it was necessary to change or move the property of
any owner or of a public utility, the property should not be moved until
ordered to do so by the engineer; and that at no time did the engineers, with
full knowledge of what was taking place, direct the power line to be de-
energized.

The supreme court said:

> We are not impressed by this argument and we are of the opinion
> that the trial court did not err in granting the motion for a judgment
> notwithstanding the verdict.
> Heslep's argument actually comes down to a contention that he
> was injured as a proximate result of the engineers' negligence in permit-
> ting the use of the crane without requiring the prime contractor to
> insulate the boom and cable against the transmission of electricity from
> the overhead electric wires and in failing to de-energize the electric
> wires in this case before permitting Heslep to attempt the removal of
> the pipe from under the power line. The contention is not sustained by
> the terms of the contracts nor by the facts in this case.
> The evidence of what happened is clear in this case. Heslep's em-
> ployer used a mobile crane in lifting heavy objects out in the street. A
> front-end loader was used to move heavy objects under, or out from
> under, overhead power lines. When Heslep was injured, he and his fellow
> employee, Faulkner, simply undertook to take a joint of sewer pipe from
> under a power line by using a crane rather than waiting for the front-
> end loader.
> The engineers' rights and powers are not to be confused with their
> obligations and duties under their contract. Without attempting to set
> out the rights, powers, obligations and duties the engineers *do have*
> under their contract, we shorten this opinion by simply holding that
> they *do not have* the right, power, obligation or the duty to supervise
> Heslep or his fellow employees in the performance of their duties. If
> Heslep's employer was in violation of any of the safety code provisions
> as to the nature or use of equipment, he should account to the proper
> authorities under the provisions of the code. The engineers who designed

and specified the materials to be used in the construction of the sewer and who assumed the responsibility of seeing that the sewer conformed to the plans and specifications were not charged with the duty of enforcing the code and were not charged with negligence as a matter of law in the employer's failure to comply with the code. . . .

In the case at bar, the engineers were not charged with the responsibility of supervising the safe removal of sewer pipe from under the power lines. As a matter of fact the contract did not provide for the placing of sewer pipe under power transmission lines or for the removal of pipe from under transmission lines at all. . . .

The judgment [in favor of the engineers] is affirmed. (*Heslep vs. Forrest and Cotton,* 449 S.W.2d 181.)

Should a Contractor Be Judged by Engineers' Standards?

"The manifest injustice of going to a contractor to design an inexpensive system and then comparing it to one which an experienced professional engineer would have designed should be apparent on its face."

This pronouncement of the Court of Appeals of Michigan, Div. 1 (an intermediate appellate court), indicates that the court was confronted by some difficult issues. In an effort to save a few dollars on a construction job, the owners and prime contractor entrusted the design of a large apartment complex heating system to the plumbing subcontractor. The system failed to perform satisfactorily, and the corporate owner of the building sued the plumbing contractor for damages. The trial court awarded the owner $9,000. The plumbing contractor, however, appealed the decision.

The court's narrative of the facts is lengthy. The essential portions follow:

Defendant Mike Gulu, doing business as Gulu Plumbing and Heating Co., is a heating contractor. On April 24, 1958, a contract was entered into by Gulu and Hampshire Home Builders, Inc., a general contractor, whereby Gulu was to design and install a heating system in an apartment house complex owned by plaintiff, Oakwood Villa Apartments, Inc. Several problems arose in the heating system and plaintiff brought this action for damages alleging that Gulu breached his contract. Trial before the court resulted in a $9,000 judgment for plaintiff and defendant takes this appeal.

Defendant began work in 1958 and the system was partially tested during the 1958–59 heating season. Pursuant to an agreed contract modification, the system utilized gas–fired boilers with attached pumps to force the heated water through the system. One boiler is located in each of the eight individual apartment buildings, which contain 100 separate apartments. Each apartment has its own hot water supply and return from the radiator and is thermostatically controlled. The flow of hot water is controlled by a motorized zone control valve which, on signal from the thermostat, automatically opens to allow the flow of hot water or closes to stop the flow.

The first problem encountered with the system was during the 1958-59 season when some of the control valves froze in the open position causing a runaway heat condition in some of the apartments. The excessive heat caused drywall damage and scorching of the walls behind the baseboard radiators requiring repairs and repainting which cost $2,239. Payments on the repair invoices were made on May 18, 1959, and July 9, 1959.

The court noted that the zone control valves were faulty. White–Rodgers Co., the valve distributor, replaced all valves in the system and gave Gulu an allowance for his labor in making the change. After the revalving was completed, the court stated, both the White–Rodgers Co. and Gulu gave their warranties on the valves from January 1, 1960, to December 31, 1960, to reassure the plaintiff and Federal Housing Administration.

The court's opinion continued:

> Because of the subsequent valve problems, which we shall discuss, plaintiff brought suit against White–Rodgers Co., who in turn filed a third–party complaint against its own manufacturer-supplier, Enco Products Corp. Both of these actions were dismissed by the court at the close of plaintiff's proofs and no appeal has been taken from this dismissal.
>
> After the valve problems developed, FHA, through William Stewart, its construction inspector, recommended withholding a portion of the contract price to insure funding of any further heating repairs and set up an escrow for this purpose. Mr. Stewart testified at trial that based on the recommendations of FHA mechanical engineers who found that the system was functioning properly, the escrow was released in January, 1960. By this date Gulu had completed the installation.
>
> On Feb. 26, 1960, preparatory to the final contract payment, Gulu sent a letter to plaintiff in which he offered a $1,000 credit against his contract balance because of the repairs made necessary by the excess heat. The proposed credit was in fact accepted and deducted from the balance due. Gulu contends here, as he did at the trial, that this credit constituted an accord and satisfaction of those damages. In its closing argument at trial, plaintiff included the drywall repairs and painting in

its itemization of damages, claiming that the credit was only for wasted fuel due to the excessive heat condition. Plaintiff's president testified that he did not appreciate the full import of Gulu's offer as made in the Feb. 26 letter.

The system operated during the 1959–60 and 1960–61 heating seasons, but testimony indicates that difficulties were experienced and complaints were received from some of the tenants. The main problems were leaky zone control valves and noise in the system.

The court's records indicate that:

. . . in March 1961, the owner retained E. G. Siegel, a consulting mechanical engineer, to inspect the system and make an evaluation and recommendations regarding any deficiencies. He prepared two detailed reports in March and April of that year. They were submitted to the owner, who included Siegel's $325 consultation fee in his damages.

Mr. Siegel made many recommendations for improving the heating system. He suggested that mounting the valves with the motor down caused any leaks to run into the motor and rust the mechanism, and the valves should be mounted in the motor-up position. The report contained recommendations for expansion joints in the pipes to alleviate the noise caused by the pipes expanding on the inflow of hot water. Additionally, he made certain findings as to desirable modifications to boiler and water controls and suggested that the boiler pumps be rotated so as to push the water rather than pull it through the heating system.

On direct examination, Mr. Siegel stated that the position of the boiler pumps did not make for a bad installation but they are usually mounted to push the system. Some of the other items referred to were stated by Mr. Siegel to be recommended but not essential. He testified that although the contract called for type "L" copper tubing, he found that defendant had installed considerable amounts of type "M" which is thinner–walled and less expensive tubing. He did state, however, that the type of tubing was only critical during the construction stage and that both types of tubing would be satisfactory in this type of system and under the pressures found in this system. When asked to express an opinion on whether Gulu had performed in a good and workmanlike manner, he declined to do so stating only that he would have included some items omitted by Gulu.

On cross–examination, Mr. Siegel indicated that his recommendations were a question of efficiency rather than functional operation; that at the time of his investigation the system was operating; and that most hot water systems made noise.

In October, 1962, the owner hired George Girk of the Girk Engineering Co. to make repairs and modifications to the heating system. The court said:

> Armed with Mr. Siegel's 1961 reports, Mr. Girk proceeded to change all 105 control valves, fix certain leaks which he attributed to faulty soldering on the installation, rewired certain transformers and thermostats, rotated the boiler pumps, and agreed further to service plaintiff's system for a period from Sept. 1, 1962, to April 30, 1963 for $40 per month.
>
> Mr. Girk's charges which were included in plaintiff's itemization of damages were: 105 new control valves, $2,262.85; installation of the new valves, $480, labor to make other repairs on the system, $389; an item referred to as labor on zone control ($174.38) which, according to the exhibits, appears to consist of $100 balance due on some controls, $65 for labor and material on transformers, and $9.38 to move a thermostat. On the $174.38 item, the $100 balance due seems to be repetitious of the cost of labor on the zone controls. Another $135 itemized by plaintiff included $105 for five new valves, which is already included in the cost for 105 valves listed above and $30 for labor in repair of leaks.
>
> White-Rodgers Co. and Enco Products Corp. moved for dismissal at the close of plaintiff's proofs. The only evidence as to the fault of the valves was that they rusted, but there was no evidence as to why they rusted. As the court stated: "There is no showing that it is a bad design, bad manufacture, improper manufacture, or faulty valves." The same finding is also applicable to Gulu in relation to plaintiff's allegations that Gulu was negligent in selecting these particular valves, and if Gulu is to be held liable for their replacement, it must depend upon his failure to install them properly.
>
> The other major item of damages was $7,575 which plaintiff contended was the cost to make all the additional corrections which Mr. Siegel recommended. At the time of trial, Feb., 1966, these modifications had not yet been undertaken.

In its opinion, the appeal court quoted the trial court's decision as follows:

> Based on the testimony that the court has heard, the court believes that the plaintiff has established its case by a preponderance of the evidence, and the question now resolves itself into the amount of damages.
>
> The court is going to award a judgment in round figures, so to speak, as best it can, because the court believes there are some things that may not have to be done to put this installation in proper working

order or to compensate for the repairs that have been made and the damages incurred as a result of the faulty installation of the system.

The system was designed by the subcontractor. He had conversations with [a representative of the owner]. However, when the conversation was reduced to writing, both sides were bound by the contract as written.

The installation was complained of continually for a period of two years at least. The work started in the heating season of '58-'59. The lawsuit was started May 10, 1962. The court finds that during that time there was no satisfactory heating system installed upon which a warranty should start. Complaints were continually made about the faulty operation and installation of the heating system.

The court, after considering all the matters and the proofs offered, and hearing arguments of counsel, is of the opinion that the plaintiff has suffered damages for the amounts that he expended and some future amounts that he should put in and should be installed in the system to make it operate according to the contract, and the court therefore awards the plaintiff damages in the amount of $9,000 plus costs.

In assessing damages at $9,000 the trial judge gave no indication of which items he allowed and which he disallowed. Normally in such a case we would simply remand to make the required findings, but in this case, based on the record made, we rule on certain items that are unallowable as a matter of law. Other items shall be referred back to the trial judge for factual determination.

The rule of damages for breach of contract is to compensate the injured party sufficiently as to put him in as good a position as he would have occupied had the breach not occurred. [Citation] Therefore, it is necessary to determine what the parties bargained for and in which respects the performance fell short of expectations.

There are two distinct, if not entirely separable provisions in this contract: to design the system and to install the system. Gulu designed the system and would be responsible for any defects in design, if they are found to render it deficient under the terms of the contract. The contract required that Gulu design a system which would meet latest I-B-R methods, FHA requirements, and inspection requirements of the city of Royal Oak, and then enumerated certain particulars to be included in the design. There is no evidence that the system failed to meet any of the above requirements. The only evidence of variation from the express items of the contract is the substitution of type "M" for type "L" tubing, but there is no testimony that substitution constituted a defect in design.

The court noted the "manifest injustice" of this situation and went on to point out that the owner could not hold the plumbing contractor to the same high standard of design if he had employed a professional engineer. It stated, "Whether this is an optimum system or not, it is all plaintiff bargained for, and all he is entitled to receive. We find no evidence which would support a finding that plaintiff received less than that to which he is entitled, and therefore we rule that the $7,575 item for proposed changes is without foundation in the record and not allowable."

Even though the court of appeals held that there was no demonstrated fault in the design and only damages for faulty installation should be recovered from the plumbing contractor, it seems fairly clear that if the professional engineers had been brought into the picture at the outset, much of this wearisome and costly litigation might have been avoided. (*Oakwood Villa Apartments Inc. vs. Gulu,* 157 N.W. 2d 816.)

Responsibility for Specifications—Where Does It End?

A Florida appellate court, the Second District Court of Appeal, made a valuable contribution to the discussion of the measure of responsibility owed by the engineer or architect to the ultimate user of the product for which he has prepared plans and specifications. In so doing, the court made some cogent remarks about the nature of professional services such as those performed by engineering firms.

The case was concerned with the alleged faulty design of roof trusses, which failed when erected. The court distinguished between negligence on the part of the designing engineers and an implied warranty of the fitness of the product for its intended use.

The plaintiff, a company which had purchased the trusses from a company which employed the defendant engineering firm, sued the defendant on two counts: first, that the defendant had been negligent in the preparation of the plans and specifications, and, second, that the plaintiff had relied on an implied warranty of fitness on the part of the defendant.

The trial court ruled that the plaintiff had no cause of action on either ground against the defendant designing engineers and the plaintiff appealed to the District Court of Appeal.

That court reversed the trial court's ruling on the alleged negligence counts but affirmed its ruling in regard to the alleged implied warranty. It stated that such a warranty applies only to the goods themselves, and not to the professional services which produced the goods.

The court's opinion follows:

This appeal arises from a judgment of dismissal determining that appellant, Audlane Lumber and Builders Supply, Inc., had no cause of action against appellee, D. E. Britt Associates Inc. The question presented is whether one who prepares the design and specifications for a chattel may be liable, upon theories of negligence or implied warranty, to a third party who is damaged by reason of defects in such design and specifications.

The case arose and was determined upon the complaint filed by appellant, a manufacturer and distributor of building supplies, against Anchor Lock of Florida, manufacturer of metal plates used in the construction of wooden trusses, and D. E. Britt Associates, Inc., an engineering firm which prepared design and specifications for wooden trusses.

The complaint alleged, *inter alia,* facts as follows: Appellant purchased a certain quantity of metal truss plates together with design and specifications for the construction of wooden trusses from appellee, Anchor Lock. The design and specifications had been prepared for Anchor Lock by appellee D. E. Britt Associates and names of both corporations were imprinted upon the plans.

Following the acquisition of the plates and plans, appellant constructed trusses according to said design and specifications and sold the trusses to a residential construction contractor. Used by the latter in the construction of a house, the trusses failed and the roof bowed, necessitating extensive replacement and repairs to the house.

This failure of the trusses, allegedly due to the fact that the design and specifications by which they were constructed were faulty, improper and unfit for their purpose, eventuated in damage to appellant's business reputation, ensuing loss of profits, and the expenditure of considerable sums in identifying and repairing the fault in the trusses.

Recovery of appellant's damages was sought under theories set forth in five counts of the complaint. Two of these counts, the fourth and fifth, are of no moment on this appeal and our consideration is limited to counts one through three. Counts one and three sound in negligence: The former against both defendant–appellees and the latter against defendant–appellee Britt Associates alone.

Alleging that Britt Associates, in preparing the design and specifications knew that they would be sold to plaintiff and others and that these purchasers would and did rely upon their accuracy and fitness, the negligence counts continued with allegations of specific acts and evidences of negligent preparation.

Count two, also containing allegation that Britt Associates knew plaintiff-appellant would purchase and rely upon the plans, alleged that Britt Associates "impliedly warranted the fitness of the design . . . for the purpose for which it was intended," that plaintiff relied upon this "warranty" and that the designs and specifications were in fact not fit for their intended purpose.

Confronted with a motion to dismiss, the lower court apparently viewed neither the negligence counts nor the implied warranty count as stating a cause of action against Britt Associates. For reasons hereinafter briefly stated, we are of the view that the court erred insofar as it dismissed the action for negligence. On the other hand, we affirm dismissal of the action on an "implied warranty."

The lower court determined that an engineering firm that prepared design and specifications for a chattel owed no duty to third persons who might be damaged by a defect in the design and that the firm's "warranty" extended only to its client. We disagree with both these determinations.

With respect to the negligence action and Appellee Britt Associates' "duty," there is no magic in the generality "professional service." The phrase "professional services" encompasses a multitude of activities which may give rise to actions on numerous theories of liability.

The nature of the professional's duty, the standard of care imposed, varies in different circumstances. So, too, the extent of the duty, the delimitation of the objects of the duty, varies. In nearly every instance duty must be defined in terms of the circumstances and the theories advanced to sustain liability.

In our view the *extent* of appellee's duty may best be defined by reference to the foreseeability of injury consequent upon breach of that duty. The complaint alleged that the appellee knew that the design and specifications it prepared would be resold to and used by various fabricators.

To argue that it is absolutely free of liability for negligence to these known users or consumers of its work is to disregard the half century of development in negligence law popularly thought to have originated in *McPherson vs. Buick Motor Co.,* 217 N.Y. 382, 111 N.E. 1050, L.R.A. 1916 F, 696 (1916) and explicitly recognized in this State in *Mathews vs. Lawnlite Co.,* Fla. 1956, 88 So.2d 299.

The allegations of the complaint bring appellant within the ambit of Britt Associates' duty and the court erred in its contrary determination.

With regard to the "implied warranty of fitness," we see no reason for application of this theory in circumstances involving professional liability. Unlike the lower court, however, we do not base our decision on the narrow grounds of privity.

An engineer, or any other so–called professional, does not "warrant" his service or the tangible evidence of his skill to be "merchantable" or "fit for an intended use." These are terms uniquely applicable to goods. Rather, in the preparation of design and specifications as the basis of construction, the engineer or architect "warrants" that he will or has exercised his skill according to a certain standard of care, that he acted reasonably and without neglect.

Breach of this "warranty" occurs if he was negligent. Accordingly, the elements of an action for negligence and for breach of the "implied warranty" are the same. The use of the term "implied warranty" in these circumstances merely introduces further confusion into an area of law where confusion abounds.

The judgment appealed insofar as it dismisses appellant's action for negligence against D. E. Britt Associates Inc., is reversed and cause remanded for further proceedings not inconsistent with the foregoing opinion and judgment.

Affirmed in part, reversed in part and remanded. (*Audlane Lumber and Builders Supply Inc. vs. D. E. Britt Associates Inc.,* 168 So.2d 333.)

This decision does not necessarily mean that the defendant designing engineers were actually negligent in preparing the plans and specifications for the roof trusses. It simply rules that the trial court erred in dismissing the allegation of negligence without a trial of the case. At the trial held as a result of the remand to the trial court, the plaintiff would have had to prove its allegations of negligence in order to prevail.

This case rejects the old rule that the engineer is responsible only to the person or company with which he has contracted directly. It adopts the more modern theory that he is responsible to the ultimate purchaser or user of the product which he has designed.

If a Wall Collapses, Who Is Responsible, Owner or Engineer?

A paper manufacturing company engaged a construction engineering firm to plan and construct an addition to one of its buildings. In the contract, both parties attempted (in the light of subsequent events, the word is used advisedly) to define precisely the areas in which each would exercise supervisory authority during the construction. This seemed a good idea at the time because each party wanted to protect its interests through full agreement with the other. But after a wall collapsed in the course of the work, owner and engineer abandoned most of their assertions of supervisory authority, and each contended that the other was practically in full charge of the operation when the collapse occurred.

The task of finally determining the actual relations between owner and engineer under the particular circumstances and in light of the language of the contract fell to the Supreme Court of Pennsylvania, in whose state the mill was located and the contract made.

The owner, Hammermill Paper Co., brought action against Rust Engineering Co. in the Court of Common Pleas, Erie County. The court granted summary judgment on the pleadings in favor of the defendant engineering company, but Hammermill appealed the ruling to the supreme court.

That court stated the situation that gave rise to the lawsuit as follows:

On March 5, 1956, Rust Engineering Co. submitted to Hammermill Paper Co. a proposal to "construct additional Finishing Facilities at [Hammermill's] Erie, Pennsylvania, plant, consisting essentially of a Finishing Extension, Roll Storage Building and Packaging Materials Building." No drawings and specifications were submitted at this time, nor was a price included in the proposal, although when executed, the contract, in this respect, provided: "When drawings and specifications are sufficiently developed, an estimate of the total cost of engineering and constructing the above facilities will be furnished [Hammermill]." Article XIII of Rust's proposal, titled "Approval after Acceptance," stated "This proposal is subject to the approval of the President of [Rust] and is not binding on [Rust] until so approved after acceptance by [Hammermill], whereupon it shall become a contract." Hammermill "accepted" the proposal on March 28, 1956, and Rust approved said acceptance on April 30, 1956.

The instant controversy involves the construction and later collapse of a brick curtain wall, 14′ 8″ high and 165′ in length, which was erected as a "vertical extension to a third story height of the existing east wall of Building No. 75." Hammermill contends the wall collapsed because of faulty and negligent construction, while Rust's pleaded position is that the cause of the collapse was an "Act of God" [excessive rains], that Hammermill had paid Rust for and accepted the work involved in constructing the original wall and in reconstructing it after collapse and was, thereby, estopped from further complaint, and finally that Hammermill has been adequately compensated by its insurance carrier for any "use and occupancy" loss incurred by the collapse of the wall. Said "use and occupancy" loss in the averaged amount of $70,392.56 arose because of the interruption of Hammermill's paper-making operation caused by the collapse of the wall through the roof of the Building No. 75 and onto the machinery in the operation and there is an additional averred loss of $11,757.53 for clean-up, maintenance and reconstruction costs involving machinery and Building No. 75, making a total of claimed loss of $82,150.09, which amount was apparently paid to Hammermill by its insurance carrier under a binder to its fire insurance policy.

Hammermill's insurance carrier in the name of Hammermill instituted this assumption against Rust in the Court of Common Pleas of Erie County to recover $82,150.09. Rust filed an answer containing new matter. Hammermill filed a reply thereto and Rust moved for judgment on the pleadings, which motion was granted. The court below, granting Rust's motion for judgment on the pleadings, concluded that Rust was . . . "An employee agency under the direct control of Hammermill" and that Hammermill was alone at fault, having retained control and responsibility for the construction of the wall. With that conclusion of the court below we do not agree.

Having thus flatly expressed its disagreement with the lower court's ruling, the supreme court set forth its reasons. The court held that Article III of the contract—Engineer-Constructor's Service—provided, with respect to the scope of work, that Rust act as Hammermill's engineering and purchasing departments, performing the work with his own forces and subletting parts when it is to the best interest of and approved by Hammermill. The specific services to be performed by Rust, and the controls and approvals to be exercised by Hammermill, were to be itemized in a job procedure to be prepared jointly by Hammermill and Rust, according to the court.

The court said that Article III also provided that Rust check all material and labor entering into the work and keep detailed accounts as necessary for proper financial management under the contract, and that the system should be satisfactory to Hammermill or to an auditor appointed by Hammermill. Most important, the court declared that the contract provided that Hammermill "shall be afforded access to the Work and to all records relating to this contract."

The court continued:

Nowhere in the agreement is there any specific language which gives Hammermill the responsibility for final approval of the design and specifications to be supplied by Rust, nor any indication that Hammermill, following the guidelines of such design and specifications, was to supervise, inspect and approve as the work progressed. In this agreement, Hammermill contracted to engage the independent services of Rust *to accomplish a particular result.* Hammermill was interested, primarily, in the result and, secondarily, in keeping an eye on the costs incurred. It is apparent from a reading of this agreement in its entirety that a certain modicum of control was agreed upon by the parties, but such control was to enable Hammermill to regulate the costs of the project.

While no hard and fast rule exists to determine whether a particular relationship is that of employer-employee or owner-independent contractor, certain guidelines have been established and certain factors are required to be taken into consideration:

"Control of manner work is to be done; responsibility for result only; terms of agreement between the parties; the nature of the work or occupation; skill required for performance; whether one employed is engaged in a distinct occupation or business; which party supplies the tools; whether the payment is by the time or by the job; whether work part of the regular business of the employer, and also the right to terminate the employment at any time." *Stepp vs. Renn,* 184 Pa. 634, 637, 135 A.2d 794, 796 (1957) and additional citation.

The fact that Hammermill retained control necessary to supervise and exercise direction over the costs feature of the work; that it secured the necessary work permits and that it retained the right to add to or subtract from the work to be done, does not convince us that Hammermill thereby occupied the status of an employer of Rust. Naturally, Hammermill was interested in the result and the cost of attaining such result but the indices of such interest as delineated in the agreement do not constitute a responsibility for the *manner* in which the work was to be done by Rust. Rust was a specialist and possessed expertise in the construction field and . . . was certainly cognizant of the application of building codes to actual jobs and the standards or strength and quality to be put into a particular job, taking into consideration the owner's needs, local surface and sub–surface conditions and the effects of weather on the resultant construction. Between the parties, Rust, *and only Rust,* had the necessary skill to perform the required work. . . . Rust, *apparently,* was to supply the tools, labor and materials to complete the work and was to be paid on a cost plus 4.2% basis for construction and erection of the required structures. Rust controlled its own workers, it assumed possession of the job site and it, necessarily, had to have control of the manner in which the work was to be done.

The court supported its position by quoting the following from a 1932 case:

The very phrase "independent contractor" implies that the contractor is independent in the manner of doing the work contracted for. How can the other party control the contractor who is engaged to do the work . . . and who presumably knows most about doing it than the man who by contract authorized him to do it? . . . If in the contract between the defendant and Kimbal, the former had reserved to himself the control of the work or had so specified the means and manner of its performance as to leave Kimbal little or no choice in the selection of the methods to be used to reach the result desired, Kimbal would not have been an independent contractor, but a mere servant or agent. . . . [Kimbal] was obedient to the will of the defendant only as to the result of the work and not as to the means of its accomplishment." *Silvens vs. Grossman,* 307 Pa. 272, 278, 281-282, 161 A. 362, 364. (1932).

The Pennsylvania Supreme Court then returned to the case at hand, stating:

The reservations of control in the case at bar, upon which the court below predicated its conclusion that an employer–employee relationship existed, were aimed toward keeping the costs of the project within bounds, toward reserving the right to expand the job if the work was found less costly than imagined, and toward reducing the cost of the work if the costs became greater than anticipated. We fail to find, on the basis of the contract between the parties, that Hammermill was insisting on retaining and reserving such a broad scope of control as to deprive Rust of the freedom of decision it would normally employ as an independent contractor and to so restrict Rust's exercise of judgment in construction matters that Rust would be considered an employee. Such a state of affairs might be shown if and when testimony was taken; however, the *pleadings* do not warrant such a conclusion so as to justify the entry of this summary judgment. . . .

The instant pleadings do not present so clearly and indubitably the existence of an employer–employee status between Hammermill and Rust as would justify the entry on that basis of a judgment on the pleadings. In that respect, we believe the court below fell into error.

The higher court added:

The court below further erred when it stated that Hammermill "does not contend that [Rust] erected the curtain wall other than as directed." Paragraph 10 (f) of the complaint alleges that, *inter alia,* Rust breached the contract "In failing to have constructed said wall in a workmanlike manner contrary to the duties assumed by [Rust] in said contract. If Rust's duties were those of an independent contractor, then Hammermill had no responsibility for nor any right to direct the manner of the construction of the wall."

Moreover, there are other issues of fact relative to the scope of Hammermill's control which arise from the pleadings. Rust pleaded that Hammermill had "examined and approved the construction plans for the project." Hammermill's answer to this was that it "approved the construction plans for the project in the sense of approving the result of the construction . . . but it specifically denied that it in any way approved or had control of the method of construction by [Rust]. " And, when Rust alleged that Hammermill had "made daily inspections of the construction site and approved said construction during the course of construction," Hammermill replied that Rust "was in complete charge of the method of performance of the terms of the construction contract and that [Hammermill's] inspections were limited to approval only to the result of [Rust's] construction activities." Such issues of fact as to the extent of Hammermill's control of the job *and* approval and/or acceptance of the result arise from the pleadings and these issues can only be resolved at trial and not at the pleading stage.

The rest of the opinion was devoted to a discussion of the right to damages on the use and occupancy issue and included several paragraphs pointing out that when the case is tried the engineering firm will have a chance to escape liability by proving to the satisfaction of the court or jury that the collapse of the wall was as it alleged an Act of God, and did not collapse as a result of faulty or negligent construction. But Rust's contention that Hammermill must bear full responsibility because of its measure of control of the job to the extent set forth in the contract was decided once for all in the supreme court's opinion. (*Hammermill Paper Co. vs. Rust Engineering Co.*, 243 A.2d 389.)

Who's Responsible for Worker Safety on a Job Site?

The precise meaning of the provision in most construction contracts regarding construction work supervision was the issue in a case decided by the Supreme Court of Arkansas.

A workman, injured by the collapse of a wall, sued the owners of the building, the architects who designed it, and the manufacturers of the concrete slabs used in construction of the wall. The trial court dismissed the case against the owners and directed verdicts in favor of the other two defendants.

The plaintiff appealed to the supreme court, which held first that the portion of the verdict dismissing the complaint against the architect be set aside and that the case be remanded for a new trial. On rehearing, however, the court changed its mind and in a divided decision ruled that the architect's supervisory duties did not require continuous work supervision of a character that would have made him responsible for detection of defects or negligence that caused the accident. The chief justice, who had written the original opinion, filed a vigorous dissent.

The opinion after rehearing began as follows:

The only issue before us is whether there was a contractual obligation upon the architect to be present continuously during construction of the funeral home where appellant, Robert Walker, a brickmason, was injured. We hold that the architect had no such contractual duty and that he had no duty to prescribe safety precautions for the contractor or to enforce performance of the safety precautions contained in the contract between the owner and the contractor to which he was not a party. . . .

The facts giving rise to this litigation show that Ruebel & Co.

employed the architect to design a funeral home on West Markham St. in Little Rock. The design for the outer walls called for precast concrete panels ten feet high, eight feet wide and three inches thick, to be backed on the inside by light aggregate blocks. This design was adopted after the architect submitted his preliminary drawings and comments to the manufacturer of the panels because of the latter's superior knowledge. and then revised the design pursuant to the manufacturer's suggested changes. Upon the architect's plans and specifications, Ruebel & Co. let the contract for construction to Cone & Stowers, appellant's employer.

At the time of the accident, appellant was laying light aggregate blocks behind the precast concrete panels. After the blocks had been laid on the east wall of the building to within two courses of the top, the braces holding the panels upright were removed to permit the top two courses to be laid. While the braces were being removed, appellant was standing on top of the wall, plumbing it. As the last brace was removed, the wall fell outward, causing the appellant's injuries. . . .

It is not contended that the architect knew the braces were being removed from the wall. The architect admittedly performed no supervisory activities in connection with the building of the funeral home. Apparently, the architect did inspect the premises from time to time.

It is the contention of appellant that the architect agreed with the owner to supervise and inspect the building, was paid a fee for it, and had a definite duty to supervise the work, including the responsibility of taking steps to secure the safety of workmen such as appellant. The architect's contention is that his duty was to supervise and inspect only to the end that when completed the building would conform to the plans and specifications and the Little Rock Building Code, and that there was no duty upon him to exercise control over means and methods adopted by the contractor which did not affect the end result, i.e., there was no duty upon him to direct or control the contractor in reference to the temporary support of the precast panels during construction.

After stating the facts regarding the contractual agreement, including a provision that the architect was to receive six percent of the contract price, of which 1½ percent was allocated to the special engineering supervision required by Section 204 of the Building Code of the City of Little Rock, the court quoted a number of that section's provisions. Those that applied directly to the matter at issue were as follows:

> *Special Engineering Supervision:* Any owner or his agent engaged in the construction or causing the erection of a building or structure where the estimated value exceeds $25,000 shall employ a registered

architect or a licensed engineer to supervise the construction of the building. Such architect or engineer shall be licensed under the laws of the State of Arkansas and *his service shall extend over all important details of framing erection and assembly and he shall render full inspection service and adequate supervision on such buildings.* [The court supplied the italics.]

He shall be held directly responsible for the enforcement of this Code, wherever same is applicable to the structure upon which he is engaged. He shall notify the Building Inspector of any attempt to conceal, patch, or repair, any defect in materials or workmanship before such materials have been examined by the Building Inspector, or his representative. He shall be held directly responsible for the infraction of any ruling of the Building Inspector, and shall have the authority to compel the removal of defective materials or to suspend or stop work, pending the ruling of the Building Inspector.

The court disposed of the building–code aspect of the issue when it stated:

> Thus it is seen there was nothing in the agreement between the architect and owner which required the architect to be continuously present during the construction of the building or to enforce the safety provisions of the contract between the owner and the contractor, unless such a provision can be found in Section 204 of the Building Code.
>
> Nor can we extend to the words "supervise" and "supervision" used in Section 204 of the Building Code, the requirement that the architect must exercise control over the means and methods adopted by the contractor which do not affect the end result. When read in its entirety, it is obvious that the purpose of Section 204 was to exact compliance with the Building Code and to hold the architect responsible to the building inspector in connection therewith.

As the General Conditions of the American Institute of Architects were specifically made a part of the contract, their effect was then considered by the court, which began by quoting Article 38 of the conditions:

> Architect's Status.
>
> The Architect shall have general supervision and direction of the work. He is the agent of the owner only to the extent provided in the Contract Documents and when in special instances he is authorized by the owner so to act, and in such instances he shall, upon request, show the contractor written authority. He has authority to stop the work whenever such stoppage may be necessary to insure the proper execution of the Contract.

As the Architect is, in the first instance, the interpreter of the conditions of the Contract and the judge of its performance, he shall side neither with the Owner nor with the Contractor, but shall use his powers under the contract to enforce its faithful performance by both.

In case of the termination of the employment of the Architect, the Owner shall appoint a capable and reputable Architect against whom the Contractor makes no reasonable objection, whose status under the contract shall be that of the former Architect; any dispute in connection with such appointment shall be subject to arbitration.

Finding nothing in the AIA statement requiring the architect to control the safety of workmen on the job, the court proceeded to quote several AIA requirements (Section 12) that made the contractor responsible for the safety of the men employed by him on the job. It then concluded:

When the architect's status under Article 38, providing that he "is the agent of the Owner only to the extent provided in the Contract Documents. . . ," is read in connection with the safety provisions of Article 12, it is at once apparent the architect was given no authority over the safety provisions except to see that the contractor designated a responsible employee whose duty was prevention of accidents and whose name and position were reported to the architect.

Construing the architect's agreement with the owner in the light of the circumstances under which it was made, the contract between the owner and the contractor, requiring the contractor to designate someone in his organization whose duty was the prevention of accidents, certainly indicates the owner was not expecting the architect to also supervise the day-to-day safety precautions, and by the fact that the contract required the name of the person so designated to be reported to the architect, and by the fact that the owner, by requiring the contractor to furnish a person to prevent accidents, had already paid once for the prevention of accidents in his building contract with Cone & Stowers.

After ruling that neither the Little Rock Building Code nor the AIA General Conditions required the architect to supervise the job continuously in such a way as to assume responsibility for the prevention of accidents, the court cited a Louisiana case (*Day vs. National United States Radiator Corp.*, 241 La. 288, 128 So. 2d 660 [1961] as further support of its decision. In the Louisiana case, the state supreme court held that the architect was not responsible for the death of a subcontractor's employee when a heating system exploded. The Arkansas court quoted, with approval, the following from that case:

As we view the matter, the primary object of this provision was to impose the duty or obligation on the architects to insure to the owner that before final acceptance of the work the building would be completed in accordance with the plans and specifications; and to insure this result the architects were to make "frequent visits to the work site" during the progress of the work. Under the contract they as architects had no duty to supervise the contractor's method of doing the work. In fact, as architects they had no power or control over the contractor's method of performing his contract, unless such power was provided for in the specifications. Their duty to the owner was to see that before final acceptance of the work the plans and specifications had been complied with, that proper materials had been used, and generally that the owner secured the building it had contracted for.

Thus we do not think that under the contract in the instant case the architects were charged with the duty or obligation to inspect the methods employed by the contractor or the subcontractor in fulfilling the contract or the subcontract.

The verdict of the Arkansas trial court in favor of the architect was affirmed, and the supreme court decided that the supervisory duties of the architect or designing engineer on a construction job does not, unless specifically provided for in the contract, include the inspection or control of measures taken for the safety of the men employed on the job. (*Walker vs. Wittenberg, Delony & Davidson Inc.*, 412 S.W. 2d 621.)

8. Bidding

Bidding procedures for both public and private construction projects provide fertile grounds for litigation. The five cases in this section are illustrations of typical disputes which arise at the bidding stage of a construction project.

Who Can Be "Lowest Responsible Bidder"?

An interesting interpretation of the phrase *lowest responsible bidder,* which occurs in almost all public construction contracts, is contained in a lengthy memorandum opinion handed down in the New York Supreme Court, Special Term, Nassau County. Admittedly, the court was confronted by a sticky situation in which some powerful forces were concerned, but it pointedly ignored the power play in reaching what it considered the just decision.

The petitioner in the matter before the court was a signal company engaged in "the business of selling, servicing, installing and maintaining signal alarms, fire alarms and other electrical equipment." The respondent was the County of Nassau.

The dispute had its inception in a transaction described by the court as follows:

> In June, 1966, the Respondent County, pursuant to a resolution duly adopted, gave notice to bidders inviting sealed proposals on Contract Number 1175-B-3, to be received by the County Executive of Nassau County on July 5, 1966, at which time the sealed proposals would be opened and read and the contract awarded as soon as practicable thereafter, for "Traffic Signal Maintenance for Area Covered by Nassau County Police District."

That prior to the date designated for the opening of bids on the foregoing bid proposal on July 5, 1966, petitioner duly executed and submitted a sealed proposal on contract documents in accordance with the notice to bidders, on the forms furnished by respondent county, and the petitioner fully complied with every requirement and specification contained in the notice to bidders, contract documents and all other documents contained in the bid proposal forms which were furnished to the petitioner by the respondent county, the performance bond in the amount required by the respondent and an experience questionnaire advising of the petitioner's qualifications.

Upon the opening of the bids it was indicated that the petitioner's bid proposal was $100,536.97; that the next lowest bid received by the respondent county was submitted by the Broadway Maintenance Corp. in the amount of $108,088.13; that the third lowest bid received by the respondent county was submitted by Littlefield Algier in the sum of $123,169.

On the morning of July 14, 1966, an officer of petitioner corporation was advised orally by a representative of the Department of Public Works of the Respondent County, a Mr. Thomas O'Donnell, that it was the successful bidder for the foregoing contract and petitioner was directed to take over the maintenance contract at 12:01 A.M. on July 15, 1966.

On the same day, namely July 14, 1966, at approximately 5:40 p.m. the petitioner received another call from a representative of the Department of Public Works of the respondent county to hold up and not to proceed for the time being, being simultaneously advised that the contract would probably start the following day. None of the foregoing facts are disputed by the respondent.

Petitioner also alleges that the Nassau County Police Department was advised by the Department of Public Works that it [petitioner] was the successful bidder under the foregoing bid proposal and would thereafter be responsible for the service and maintenance of its traffic signal system, because the president of petitioner, on the morning of July 15, 1966, had received two telephone calls from two police precincts in Nassau County requesting immediate traffic light repairs by petitioner.

That the petitioner must have been designated by the responsible officials in the Department of Public Works of the respondent county although a formal contract had not yet been signed, is further evidenced from the fact that petitioner was advised on July 14, 1966, that it was the lowest responsible bidder and service was to start at midnight of the next day, which obviously permitted insufficient time within which to draw and execute a formal contract.

Subsequently and for approximately twelve days thereafter

petitioner and its attorney were unable to obtain definite word from the respondent's officials, including the Department of Public Works or the Legal Department as to what action the respondent contemplated taking, being delayed from day to day. Late in the day of July 26, 1966, petitioner received a notice of rejection in writing over the signature of Daniel T. Sweeney, Deputy County Executive, advising that the respondent did not consider the petitioner "to be the lowest responsible bidder and the awarding of the contract to you would not be in the best interest of the county," also that "it is our intention to award the contract to the lowest responsible bidder."

The court then stated in considerable detail the circumstances that had caused this about-face on the part of the respondent county. Apparently a communication had been received from an officer of a local labor union objecting to the award of the contract to the petitioner. The court quoted from the union official's letter, letter saying he "vehemently protests an award of this contract by the County of Nassau to Long Island Signal, which is a non-union concern," and also from another communication stating that an apprentice was employed by the petitioner and was not paid the prevailing wage for his work.

In defense of the county's action in refusing to award the contract to the petitioner, the County Attorney relied chiefly on a paragraph of the contract K-1 that read as follows:

> Prevention of Delay
> The contractor and his subcontractors shall not employ on the work any labor, materials or means whose employment or utilization during the course of this contract, may tend to or in any way cause or result in strikes, work stoppages, delays, suspension of work or similar troubles by workmen employed by the contractor or his subcontractors, or by any of the trades working in or about the buildings and premises where work is being performed under this contract, or by other contractors or their subcontractors pursuant to other contracts, or on any other building or premises owned or operated by the County of Nassau. Any violation by the contractor of this requirement may, upon certification of the Commissioner of Public Works, be considered as proper and sufficient cause for cancelling and terminating this contract.

A second affirmative defense was asserted under a provision of the Labor Law.

The court's reply to these defenses reads as follows:

> The court has already dealt with the Second Affirmative Defense under the Labor Law and will now discuss that part of respondent's

defense which concerns itself with Paragraph "K-1" of the proposed contract, since the County Attorney as legal advisor to the County Executive and the Department of Public Works of Nassau County has advised these officials that in his opinion they would have the right to terminate the contract if it was awarded to the petitioner after petitioner started to work thereunder and if and when the labor unions carried out their threat to bring about a work stoppage (as will be hereinafter dealt with), that therefore, in anticipation of strikes or work stoppages which would be contrary to the interest of the County of Nassau, the respondent county would be justified in rejecting petitioner's bid at this time rather than await subsequent trouble.

Our laws, both State and Federal, permit non-union as well as union shops to bid on public works contracts. This is inherent in our existing statutes, as is the right of every citizen to make his own determination as to whether or not he desires to become a member of a labor union.

If the court were to endorse the interpretation based on the aforequoted Paragraph "K-1" of the proposed contract that the respondent seeks to place thereon, it would vitiate completely any right that a non-union shop now has to bid on a public works project, for then every department of government, municipal, state or federal could justify the rejection of every bid submitted by a non-union shop no matter how much lower the bid is than that of other bidders, on the premise that if they awarded the contract to the lowest responsible bidder they would be justified in terminating the contract as soon as work got underway if they received a threat from any labor organization of a work stoppage on other jobs. This is hardly what our federal and state Constitutions contemplated nor does existing statutory or decisional law so hold.

The court quoted from a letter from the Commissioner of Public Works to the County Executive recommending the rejection of the petitioner's bid. This paragraph was stressed, "The above entitled contract has a clause K-1, 'Prevention of Delay.' This clause gives us the authority to cancel a contract should it cause labor problems."

Commenting on this assertion of authority the court said:

It is abundantly clear, that, based upon the coercion, intimidation and threats . . . the responsible officials of the respondent county felt it was perhaps the better part of discretion to award this contract to a higher bidder than the petitioner rather than face the wrath of labor union officials in their county, even though the actions of the latter would be illegal under existing law.

This court disagrees with the respondent's County Attorney in his reasoning that the clause would give the county authority to terminate a contract for a breach thereof can be anticipated to the point where a lowest responsible bidder is rejected on the theory that there may be a breach in the future after the contract is in full force and effect which might give the county cause for termination.

The court then reached the following conclusion:

Section 2206 of the County Government Law of Nassau County, Laws 1936.c.879, taken from Section 103 of the General Municipal Law of the State of New York, clearly sets forth the terms and conditions under which contracts on public works projects must be awarded, provided the bidder meets with all of the requirements therein set forth, the applicable language being "the contract shall be let to the lowest responsible bidder. . . ."

That the petitioner is the "lowest responsible bidder" in this case has been overwhelmingly established, and the court concludes that under all the circumstances involved herein petitioner has met the requirements of existing law and that the county would not be justified and that it would not be in the best interests of the public welfare to award this contract to anyone other than the "lowest responsible bidder," which in this instance is the petitioner.

Accordingly the court directs that judgment be entered herein directing the public official or officials of the Respondent County of Nassau who are authorized to make the award of Contract No. 1175–B-3 to make such an award to the petitioner herein. (*Long Island Signal Corp. vs. County of Nassau.* 273 N.Y.S.2d 188.)

Is It Possible to Get Out of a Low Bid?

If a person, firm or corporation submits a sealed bid on public works, the principle contended for by the contractor, namely, that after all the bids are opened he can withdraw his bid under the plea of a clerical mistake, would seriously undermine and make the requirement or system of sealed bids a mockery; it could likewise open wide the door to fraud and collusion between contractors and/or between contractors and the Public Authority.

What is the use or purpose of a sealed bid if the bidder does not have to be bound by what he submits under seal? What is the use or purpose of requiring a surety bond as further protection for the public, i.e., the municipality, if a bidder can withdraw his bid under plea of clerical mistake, whenever he sees that his bid is so low that he must have made an error of judgment?

These words, quoted with approval by the Supreme Court of Pennsylvania from a 1958 decision it made, emphasizes, via the rhetorical question, the importance of adhering strictly to the full requirements for sealed bids. The sentences quoted were part of an opinion handed down by that court in a case involving an attempt to withdraw a bid on a school building construction job.

The facts of the situation were stated by the court:

The Public School Building Authority advertised for bids for the construction of the Baden Economy Joint School District in Beaver County. The advertisement provided the instructions and conditions of the job and, in addition, provided the manner of submitting the bid and the manner of withdrawal of the bid. The instructions provided that the bidder or his agent would personally appear at the authority office with a written request to withdraw prior to the time set for the opening of the bids.

On October 4, 1960, appellant, Modany, had completed the preparation of his bid. At 5:10 p.m. on October 4, 1960, Albert Modany, the son and agent of Gabriel Modany and the person responsible for the preparation of the bid, delivered the bid, together with a bid bond in the penal sum of $13,000, to the architect for the job, at his office in Ellwood City, with instructions to deliver the bid to the Authority's office in Harrisburg. This the architect did, at 9:45 a.m., October 5, 1960.

Sometime during the evening of October 4, after the bid had been delivered to the architect, Albert Modany discovered that he had omitted an approximate $120,000 item from the bid. Albert Modany, then being unable to contact Mr. McCandless, the architect, by telephone, then made no further effort to withdraw the bid until about three hours later when he returned from visiting his wife in the hospital. He then sent a telegram to the authority at its main office in Harrisburg to the attention of the architect.

This telegram was sent at 9:30 p.m. on October 4, 1960. The telegram stated that the bid was to be withdrawn, but did not mention that a mistake had been made. At 10:05 a.m. on October 5, 1960, 25 minutes prior to the opening of the bids, the contents of the telegram were made known to the counsel for the Authority, the Executive Director of the Authority, and an employee of the Authority authorized to receive bids. At 10:30 a.m. October 5, 1960, all of the bids were opened. The appellant's bid was $211,700. The next lowest bids were

$342,800 and $346,495. The appellant's bid being the lowest, his bid was accepted.

The appellant then sought to enjoin the Authority from acting on his bid and to have the bid declared null and void, because it had been withdrawn. The lower court had before it for consideration the effectiveness of the withdrawal of Modany's bid and whether rescission of the contract should be decreed if the bid withdrawal was not effective. The court denied relief and dismissed the complaint and that adjudication was affirmed by the court *en banc.* This appeal followed.

The court then stated the provisions of the Instructions to Bidders relating to withdrawals of bids. They read as follows:

WITHDRAWALS OF PROPOSALS

13. Bidders will be given permission to withdraw any proposal after it has been received by the State Public School Building Authority, provided the bidder, or his agent duly authorized to act for him, personally appears at the office of the State Public School Building Authority with a written request prior to the time set for the opening of bids. At the time set for the opening of proposals the withdrawn proposal will be returned to the bidder. Such proposal will not be publicly read at the bid opening.

FAILURE TO EXECUTE CONTRACTS

21. If the lowest responsible bidder to whom the contract is awarded fails to give bonds or execute the contract within the time specified in the proposal, the amount of the proposal guaranty shall be forfeited to the State Public School Building Authority, not as a penalty but as liquidated damages.

The court then pointed out that the bidder's conduct in the case before it did not comply with these conditions:

Modany's telegram read: "Reference Contract A10-675-1 request withdrawal of our bid." There was no other effort made by Modany to withdraw his bid nor to contact the Authority or the architect in an effort to withdraw his bid.

The telegram gave no indication that a mistake had been made in the bid. Although Mr. McCandless, the architect, upon receiving the telegram, attempted to contact Modany by telephone, he was unsuccessful, as no one answered the phone at the Modany office prior to the opening of the bids. When the bids were opened, the Authority had no knowledge at that time that the appellant, Modany, had made a mistake in his bid.

Modany had knowledge of the requirement for the withdrawal of proposals and was bound to follow the requirements in order to accomplish a withdrawal of the proposal. He did nothing except to send the unverified telegram which did not mention a mistake in the bid. Instructions to Bidders are a material part of a contract. The Instructions to Bidders were part of the contract documents in this case. . . .

The orderly procedure of making an offer, particularly in regard to public contracts, requires care that the bidders be protected as well as the public interest. This care extends to the procedure in withdrawal of bids to prevent mischief. Reasonable regulations that protect these interests and are consonant with contract law will be upheld. An unverified telegram, and nothing more, as a means of withdrawing a bid is not conducive to the protection of bidders competing for public business. Reasonable and fair provisions surrounding competition for contracts for the performance of public works is necessary and, therefore, they must be established and maintained.

The record indicates that Modany failed to withdraw his proposal in the manner provided for in the Instructions to Bidders. There seems to have been no attempt (except the sending of the unverified telegram without a reason stated therein) to comply in any manner with the withdrawal provision. The indifference shown by the record, reveals an attitude of complete unconcern. It would be difficult to conceive a more negligent disregard for the requirement of withdrawal of a bid.

After quoting at length from the Chancellor's opinion in the court below, in which it was pointed out repeatedly that Modany had neglected his opportunities to comply with both the letter and spirit of the law, the court summed things up with the laconic statement, "A public business should not be conducted in this fashion."

The court returned briefly to the Chancellor's opinion in the trial court saying, "The Chancellor aptly said: 'For the general welfare of the public this type of loose conduct in the contractual relations arising from public contracts should not be condoned. To so condone it would not only be placing a premium on negligence, but would be opening the door for fraudulent conduct between bidders or between a bidder and the public body inviting the bid. For these reasons strict compliance with the Instructions to Bidders must be insisted upon.'

"We agree that Modany fell far short of effectively withdrawing his bid." (*Modany vs. State Public School Building Authority*, 208 A, 2d 276.)

Can Public Project Bids Be Readvertised?

A New York statute governing competitive bidding on public projects was interpreted by the Supreme Court, Appellate Term, Queens County, in an opinion of interest and importance to bidders on such work. The New York City Board of Education was the respondent in an action brought by a low bidder on one phase of a contract on which separate bids for various portions of the work had been invited. All bids were rejected because their total exceeded the estimated total for the whole job, and new bids were invited for the job as a unit. The contractor who had submitted the low bid for the electrical work on the reconstruction of the Far Rockaway High School Athletic Field, including appurtenant structures, contended that the statute forbade such action by the board.

The board's Superintendent of Design prepared separate specifications for the work and advertised for bids. When they were opened they were found to be as follows: Item 1—General Construction, $184,000; Item 2—Plumbing and drainage, $12,000; Item 3—Heating and ventilating, $1,000; Item 4—Electrical work and lighting fixtures, $3,000. The aggregate of $200,000 exceeded the estimated cost by $40,000, and the Superintendent rejected all bids and readvertised for bids.

The new advertisement, however, did not include separate specifications but requested a single bid "for all of the work, which shall include the cost of the work of general construction, plumbing and drainage, heating and ventilating, and electrical work and lighting fixtures; any statement or information to the contrary in the specifications or on the drawings shall be disregarded."

This sweeping disregard of the specifications was challenged by the electrical contractor. The Association of Contracting Plumbers of the City of New York was granted leave to appear throughout the proceedings as *amicus curiae,* because of the association's direct interest in the outcome of the action.

The statute governing bidding on municipal projects, which the plaintiff contended had been violated, reads as follows:

Separate specifications for certain public work

(1) Every office, board or agency of a political subdivision or of any district therein, charged with the duty of preparing specifications or awarding or entering into contracts for the erection, construction, reconstruction or alteration of buildings, when the entire cost of such work shall exceed $50,000, shall prepare separate specifications for the following three subdivisions of the work to be performed:

a. Plumbing and gas fitting;

b. Steam heating, hot water heating, ventilating and air conditioning apparatus; and

c. Electrical wiring and standard illuminating fixtures.

(2) Such specifications shall be drawn so as to permit separate and independent bidding upon each of the above three subdivisions of work. All contracts awarded by any political subdivision or by an

officer, board or agency thereof, or of any district therein, for the erection, construction, reconstruction or alterations of buildings, or any part thereof, shall award the three subdivisions of the above specified work separately in the manner provided for in section one hundred three of the charter.

The court described the work to be done as consisting of "the demolition and removal of existing bleachers, construction of new bleachers, construction of a toilet and storage building beneath the bleachers, alterations to an existing locker room and handball wall, construction of a football field and athletic track and jumping pits. . . ." The court stated the contentions of the parties as follows:

Petitioner contends that the readvertisement for the project based upon a single bid violated section 101 of the General Municipal Law since separate specifications were not prepared in accordance therewith. Respondent contends that (1) section 101 only applied to construction which requires all three types of work specified in the three subdivisions enumerated in said section and (2) since the only building involved in the project is the toilet and storage building, and since its cost does not exceed $50,000, compliance with section 101 is not required.

The court first took up the board's contention that all three types of work enumerated in the statute must be included in the work if the requirement for separate specification and bids is to apply. On this subject the court said:

It is not necessary that all the work enumerated in the three subdivisions of Section 101 be part of a project in order that it be governed by said section. Those subdivisions merely describe, in general terms, the type of work which must be separately classified solely to the extent that such work is to be performed. It can hardly be suggested that a project would not be governed by Section 101 if the work to be performed did not include some part of the work enumerated in any one of these three subdivisions or if the project required some work in addition to that specifically enumerated in said subdivisions.

The court thus ruled that the action of the superintendent in readvertising for a single bid on the entire project violated the statute.

The second point raised by the board, namely that as the only "building" involved in the project was a small structure costing much less than $50,000, thus making the statute unenforceable so far as this particular project was concerned, was then discussed by the court:

Section 641-1.0 of chapter 26 of the Administrative Code of the City of New York relating to the Department of Buildings defines the word "building" as follows:

"The term 'Structure' shall mean a building or construction of any kind."

In light of these definitions, it is the opinion of the court that the entire project, *or any one of the following classifications of work,* would constitute the "erection, construction, reconstruction or alteration of buildings" under Section 101, would cost more than $50,000, and would accordingly require preparation of separate specifications:

(1) Removal and construction of bleachers, construction of the toilet and storage building and alterations to the locker room and handball wall.
(2) Construction of the bleachers and the storage building and alterations to the locker room and handball wall.
(3) Construction of the bleachers and the toilet and storage building.
(4) Construction of the bleachers and alterations to the locker room and handball wall.
(5) Construction of the bleachers.

Having made this broadest possible application of the word *building* and again ruling in favor of the petitioner's demand for separate bids on the various phases of the project, the court made one concession to the board. It ruled that in view of the fact that the mayor's approval had mentioned the specific sum of $200,000 for the entire job, the board's action in rejecting bids which exceeded that amount by $40,000, was not arbitrary or capricious, thus leaving the board free to readvertise for bids, with the express understanding that separate bids would be called for on the various phases of the work as required by the statute. (*Reynolds Electric Co. vs. Board of Education,* 259 N.Y.S. 2d 503.)

When Can a Bidding Error Be Rescinded?

Withdrawal or rescission of a low bid on a construction project is difficult, but it can be done under certain circumstances. This was demonstrated in a case decided by the Superior Court of New Jersey, Chancery Div.

The Chancery Div. is a court of equity, which is an inheritance from the ancient days in England when the common law courts could give relief only on specified common law writs. Other obvious wrongs were referred to the King's Council and later to the head of the Council, the Chancellor.

Most modern courts have equitable jurisdiction over actions in which it is impossible to measure the relief in terms of monetary damages, as in the

New Jersey case. Courts of equity traditionally have a wider discretion than courts of law; in fact, the phrase *the conscience of the court* has been applied to them. In the exercise of this discretion, they frequently take into consideration facts and circumstances, as in this case, which seem to vary the terms of the original agreement.

The case arose out of competitive bidding on public construction work. Cataldo Construction Co., a general contractor in Orange, submitted on May 22, 1969, in response to advertising by Essex County, a bid of $24,233 for work to be done on the Lower Chatham Bridge, which spans the Passaic River between Morris and Essex counties. To its bid Cataldo attached a certified deposit check for $2,500, as required by the instructions to bidders.

When the bids were opened on May 22, the Cataldo bid was lowest. A resolution of the board of freeholders of Essex formally approved the award of the contract to Cataldo that day. Although the advertisements soliciting bids made no mention of any active participation in the project by Morris County, the Essex resolution was expressly made contingent on a concurring resolution by the board of freeholders of Morris County. Such a concurring resolution was adopted in Morris on May 28.

Cataldo's bid had been prepared hurriedly to meet the advertised deadline. As a result of the haste, a mistake was made in the figure submitted. One of the amounts transferred from Cataldo's work sheet to its sheet for totals was listed at $13,822, rather than the $23,822 it should have been. The mistake was carried over to the typewritten proposal.

The $10,000 mistake was discovered the day after the bids were opened and the Essex resolution adopted but five days before the adoption of the Morris resolution. According to the court, no proof was offered that Cataldo knew at this stage that the Essex resolution was contingent on concurrence by the freeholders of Morris County or that efforts were made to notify the proper parties in Morris before the adoption of that county's resolution.

The court said:

> It is clear, however, that immediately upon the discovery of the error, Cataldo's president telephoned the Essex County engineer to explain the mistake. A few days later, apparently before the Morris resolution was adopted, Cataldo's president met with the Essex engineer, showed him the work sheets, and was told nothing could be done to correct the error. By letter dated May 28, 1969, Cataldo again gave Essex County notice of the error and requested that its bid be withdrawn.

Essex County refused to rescind the bid and return Cataldo's deposit check. Cataldo commenced action against Essex County, the sole defendant. Both parties treated the dispute as a matter solely between themselves and not involving Morris County.

Cross–motions for summary judgment came before the court. Cataldo sought rescission of the bid and return of the deposit; the county asserted that the bid should not be rescinded and that it was entitled to retain the deposit as a forfeiture for Cataldo's failure to sign a contract and perform the work for $24,233. There were no issues of fact; the county's brief admitted all of the facts as stated by Cataldo. The court noted that:

> The law is clear that a competitive bid is an option based upon a valuable consideration, namely a privilege or bidding and the legal assurance to the successful bidder of an award as against all competitors. [Citations] As such, the bid is both an offer and a unilateral contract; when it is accepted, it becomes a mutually binding contract. [Citations] So here the acceptance of Cataldo's bid by the appropriate governing bodies of Essex and Morris Counties created a valid contract.

The question became, then, whether Cataldo was to be relieved in equity from the obligations of its contract. The sole ground asserted for relief was Cataldo's unilateral mistake in the computation of its bid. The court stated:

> It is the general rule that a unilateral mistake of fact, unknown to the other party, is not ordinarily ground for avoidance or rescission. Nevertheless, there can be no question but that equity, under appropriate circumstances, may grant relief by way of rescission of a unilateral mistake of fact. [Citations] To qualify for the equitable relief sought, Cataldo must show special circumstances justifying a departure from the generally controlling principle that parties are bound by the contracts they make for themselves.

The leading case in New Jersey on the question at hand is *Conduit & Foundation Corp. vs. Atlantic City*, 2 N. J. Supra 433, 438, 64 A.2d 382 (Ch. Div. 1949). There, too, the plaintiff erred in computing a bid for performance of construction work, and the resultant figure was substantially lower than it would otherwise have been. Although the city was informed of the error before it accepted the bid, it refused to allow the attempted withdrawal of the bid and insisted on its right of acceptance. The plaintiff sued for equitable relief on the ground of unilateral mistake of fact. The case was tried before Judge (now Justice) Haneman, then sitting as a superior court judge in chancery. In his reported opinion, Judge Haneman set forth the criteria for relief as follows:

> The essential conditions for such relief by way of rescission for mistake are (1) the mistake must be of so great a consequence that to enforce the contract as actually made would be unconscionable; (2) the

matter as to which the mistake was made must relate to the material feature of the contract; (3) the mistake must have occurred notwithstanding the exercise of reasonable care by the party making the mistake; and (4) it must be able to get relief by way of rescission without serious prejudice to the other party, except by loss of his bargain.

Judge Haneman found that under the circumstances before him each of those conditions had been met. That being the case, he allowed the plaintiff to rescind the bid and recover the accompanying deposit.

The judge in the Cataldo case said:

> Putting aside for the moment the question of Cataldo's care in the preparation of its bid (the third criterion listed above), it appears that Cataldo is otherwise entitled to the relief sought.
>
> The mistake reduced Cataldo's bid from $34,233 to $24,233—a substantial margin of error. Moreover, Essex County's refusal to allow Cataldo to revoke its bid upon discovery and immediate notification of the error—and before the necessary acceptance of the bid by Morris County—lends additional strength to the conclusion that to enforce the contract here as actually made would be unconscionable. As Judge Haneman said in *Conduit & Foundation Corp., supra:*
>
> "Not only was the mistake of so great a consequence, but the defendant's conduct as well was such as would make an enforcement of the contract unconscionable."
>
> The matter to which the mistake related—the price—was obviously a material feature of the contract.
>
> Although the remedy of rescission for unilateral mistake ordinarily is only available so long as the *status quo ante* can be restored [Citations], there is no indication here that Essex County would be, or has been, seriously prejudiced by rescission, except for the loss of its bargain. It is true that in *Conduit & Foundation Corp., supra* the mistake was discovered and revocation of the bid attempted *prior* to its acceptance by the contracting governmental body, with Judge Haneman proclaiming that "under the facts here present such bid was properly rescinded within time." [Citation] However, in *Barlow vs. Jones, supra,* an earlier case cited with approval in the *Conduit & Foundation Corp.* opinion, a contractor was allowed rescission of his bid and return of his deposit on the ground of his own unilateral mistake of fact, despite the fact that the error there was discovered *after* the bid had already been accepted. Vice–Chancellor Reed in *Barlow* pointed out that there had been no prejudice to defendants such as would bar plaintiff's right to relief.

In the *Barlow* case, the court said, notice was promptly given to the commission of the existence of the mistake and of Barlow's inability to execute the contract. The commission then accepted the bid of the next highest bidder just as it would have done had Barlow never bid. There had been no change in the status of the parties included by the belief that Barlow would execute his contract or by any negligence in promptly reporting the mistake. The court continued:

> Here Cataldo discovered its error, and called the attention of Essex County to it before the acceptance of Morris County and at a time when no serious prejudice had been suffered by Essex County. Cataldo's claim for equitable relief cannot be defeated on this basis.

> So we come finally to the troublesome issue of whether Cataldo exercised reasonable care in the preparation of its bid. As was indicated in the preceding discussion of *Conduit & Foundation Corp.*, equitable relief is not available for unilateral mistakes of fact caused by a failure to exercise reasonable care. [Citations] The question then becomes, what is "reasonable"? Whether the negligence of the mistaken party is of such a degree as to bar equitable relief depends in each case on the particular circumstances.

The judge said that in *Crane, supra* the court made these comments:

> There are many instances in which the equity courts have in their discretion refused relief where there was negligence on the part of him by whom the doctrine of mistake was invoked. . . . But in *Murray vs. D'Orsi*, 98 N.J. Eq. 548.131 A. 122.123 (Ch. 1925), the court observed that "a party's own negligence will not always occasion refusal of relief . . . each instance of negligence must depend to a great extent upon its own circumstances," and that in many instances "the complaining party has been relieved of the consequences of his mistake of fact, even where it was due to his own clear negligence."

Mistake by its very definition, the court-of-equity judge said, implies some degree of negligence. Human failing is its essence, and it denotes error of judgment; however, it still remains the obligation of a court of equity to determine whether, despite such misjudgment, it would be inequitable and fundamentally unjust not to set aside the sale. The judge continued:

> Cataldo does not have proofs to excuse its mistake as strong as the evidence offered by the contractors in *Conduit & Foundation Corp.* and in *Barlow vs. Jones, supra,* to excuse their respective errors. In

Conduit & Foundation Corp. it was shown that work on the contractor's bid was quite complex and time consuming, and that on the whole it had been carefully done. . . .

And in *Barlow vs. Jones, supra,* the proofs indicated that the contractor was a sick man who broke down during the preparation of his bid and was unable to continue; the error in that case was produced by the contractor's bookkeeper, who compassionately and hurriedly completed the bid. Vice-Chancellor Reed decided under the circumstances that "the mistake was one which . . . might occur to a careful man and was certainly one which cannot be characterized as grossly negligent." [Citation]

In the present case the affidavits of Anthony G. Cataldo and his estimator, Carr, state that Carr got his instructions to prepare a bid about one week before the bidding date; that Carr met with delays in getting quotations on the prices of some materials needed for the work; that it was not until about ten o'clock in the morning of the bidding date that he was able to obtain final figures, which, after a brief conference with Cataldo, were typed hurriedly on the bid form prescribed by the county. Then the bid was taken by automobile from the office of the Cataldo company to the Hall of Records just in time for delivery at the appointed hour of eleven o'clock. Carr's affidavit reads in part:

"I gave the work sheets to the secretary with instructions to quickly prepare the typewritten bid. In my haste I transposed the sum of $13,822 from my working cost sheet P-1, instead of the true sum of $23,822 to my total sheet P-3-A, which shows a total of $24,233 instead of $34,233."

Though pressure of time is the only real excuse, or explanation, for plaintiff's error, my conclusion is that relief should be granted. Discovery of the mistake and the giving of notice were timely. Because of the delay in getting price quotations on materials the final bid figures were worked out and put in form for submission in an atmosphere of great haste. In that atmosphere the recording of an incorrect figure seems normal enough to be excusable. Loss of all benefits of a bargain based on the mistake appears to be the only disadvantage to the county which will flow from the granting of relief. At oral argument I was furnished with the following information. When Cataldo gave notice of its mistake and refused to sign a contract at $24,233, new bids were advertised for promptly and resulted in a low bid by Cataldo at $34,233 (or substantially that amount), that being in effect a corrected version of Cataldo's original bid and a figure lower than any bid received from any other bidder at the original bid opening. On this second Cataldo bid a contract was executed and in due course fully performed.

The annotation already mentioned (52 A.L.R. 2d 792) shows that the current of decisions in a substantial number of jurisdictions has been strongly in favor of relief for the erring contractor in situations where a mistake is promptly discovered and called to the attention of the public body before the execution of a formal contract and there is no substantial disadvantage to the solicitor of bids other than loss of a bargain based on the bidder's error.

The judge rescinded Cataldo's bid and ordered Essex County to return the deposit of $2,500 made when the bid was submitted. (*Cataldo Construction Co. vs. County of Essex,* 265 A. 2d 842.)

The Low Bid—a Grievous Mistake?

Every contractor experiences a thrill of satisfaction when word comes through that his bid is low on an important construction job. But every now and then that thrill of satisfaction turns to dismay when he discovers that the reason he is the low bidder is the discouraging fact that a grievous mistake has been made.

Perhaps he makes a mathematical error in quantities or prices, or as in the case in the Maryland contract discussed here, he makes a mistake in the interpretation of the bidding documents. Such an error may easily wipe out all anticipated profits, and in some cases may lead to bankruptcy.

The Maryland case had its beginnings when a college in that state decided in 1961 to construct a student union and a new dormitory on its campus. Through its architects, the college issued an invitation for bids. Under a loan agreement with the Federal Housing and Home Finance Agency, the college was obligated to contract for the work upon free, open, and competitive bidding, and to award each contract after approval by HHFA to the lowest responsible bidder as soon as practicable.

Action was brought in the United States District Court for the District of Maryland when the successful bidder misinterpreted the terms of the contract. The bidding requirements were stated as follows:

The bidding documents, which contained the usual general provisions, provided for three base proposals: (A) Combined bid for the construction of the Student Union Building and the Dormitory Building; (B) Student Union Building only; and (C) Dormitory Building only. The college reserved the right to reject any or all bids, and if the lowest bid submitted by a responsible bidder exceeded the amount of funds available to finance the contract, to reject all bids or award the contract on the base bid modified by certain deductible alternates. The Instructions to Bidders provided that any bid might be withdrawn prior to the scheduled time for the opening of bids, but that no bidder might with-draw a bid within thirty days after the date of opening.

Each bidder was required to post a bid security bond in the amount of 10 percent of its bid, and the bidding documents provided that if a successful bidder should fail or refuse to execute and deliver the contract and bond within ten days after receipt of written notice of acceptance of its bid, the bid security should be forfeited to the college as liquidated damages.

When the successful bidder, Miller Inc., discovered its mistake and sought cancellation of its bid and return of its bid bond, the college brought an action against the surety company which had provided the bond, to re-cover the penalty required by the provision quoted in the previous paragraph. The contractor intervened as a defendant. The suit was brought in the federal courts because of a diversity of citizenship among the original parties. The amount in dispute was requisite to confer jurisdiction.

Miller Inc. was one of nine contractors who requested permission of the architects to submit bids and received complete sets of bidding documents including the plans and specifications. Six of them submitted bids.

When the bids were opened, Miller's combined bid was low and amounted to $1,389,450. The next lowest bid was $1,478,064. There was one bid lower than Miller's separate bid on the dormitory.

The difference of nearly $90,000 between Miller's bid and the next lowest bid was soon explained. The bidding documents called for separate price quotes on certain items, including kitchen equipment and snack bar for the student union building, built-in furniture for the dormitory, and certain plumbing and electrical work. All bidders except Miller had included these items in their base bids. As Miller's estimate of the cost of these items was $160,555, the misinterpretation of the bidding requirements made a sub-stantial difference.

The court stated the situation:

> Both W. Harley Miller, president and principal stockholder of Miller Inc., and Jenkins, its estimator, interpreted and construed the Form of Proposal as not calling for the inclusion of the Separate Price Quotes in the Base Proposals, and Miller Inc. did not include any of these Separate Price Quotes in its Base Proposal for the construction of the buildings.
>
> The architects and the college intended the form of Bid Proposal and the Specifications to mean that the kitchen equipment and the built-in furniture, for which Separate Price Quotes were requested, should be included in the Base Bids for the Student Union Building and for the Dormitory Building, respectively, and in the Combined Base Bid

for both buildings. All of the bidders except Miller Inc. so interpreted the bidding documents and considered that these special equipment items were included in their Base Bids; all of their bids were prepared and submitted on that basis.

After reciting other details and stating that Miller and Jenkins had relied on practices in the Baltimore building industry, the court continued:

> The Court does not find that such reliance was justified. The Form of Proposal prepared by the architects was confusing, but it was not uncertain or ambiguous in the sense that it was subject to more than one reasonable meaning. The only reasonable meaning was the one intended by the architects and the college and recognized by all of the other bidders, namely, that the special equipment items for which Separate Price Quotes were requested, should be included in the base price. . . .
>
> In omitting the special equipment from its Base Bids, Miller Inc. did not make a clerical, mechanical or mathematical error. The bids which were submitted were made deliberately, intentionally and not inadvertently. The omission was an honest mistake or misunderstanding, made in good faith, and resulted from Miller's interpretation of the Form of Proposal as meaning that the Separate Price Quotes were not part of the Base Proposals.
>
> The mistake was substantial and material. The omission amounted to $169,000 if based upon the Separate Price Quotes of Miller Inc. for these items, $160,555 if based upon the lowest quotations which Miller Inc. received from suppliers. The omission was more than the $140,000 which Miller Inc. had included in its Base Proposal for overhead, bond expense and profit.
>
> Miller Inc. was a responsible bidder. It was an experienced contractor and was then engaged in four or five construction jobs. It was financially able to perform its contract with the college, although if it had done so, it would have suffered a loss.
>
> To a person experienced in the building industry, such as the other bidders or the architects, it was apparent that Miller's bid was probably the result of some error or mistake. It was $89,000 lower than the next lowest bid, and the next four were within a spread of $20,000.
>
> On the other hand, the representatives of the college, other than the architects, did not realize that there was a material error in the bid.

At this point in the court's lengthy opinion, things did not look very promising for the unfortunate contractor despite statements that he had made an honest mistake in good faith.

But the court had some Maryland law up its judicial sleeve that enabled the erring contractor to get off the hook. After a recital of attempts on the part of the college to force a quick start of the work in order to avoid increased labor costs, and the unsuccessful attempts of Miller to obtain cancellation of the contract, the court said, in part:

Since jurisdiction of this action is based on diversity of citizenship, the substantive law applicable to this case is the law of Maryland, where the contract was made and to be performed. Procedural questions are governed by Federal law.

No case precisely in point decided by the Court of Appeals of Maryland, or by any other court, has been cited or found. But the general principles which must control this case are set out and discussed in *Baltimore vs. DeLuca-Davis Construction Co.*, 210 Md. 518, 124 A.2d 557 (1956).

In that case DeLuca-Davis had submitted to the City a bid of $1,796,064.25, which by reason of a clerical, material and palpable error, made in good faith, was at least $589,880 less than it was intended to be and some $700,000 less than the engineer's estimate and the next lowest bid. Before the bid was accepted by the City, DeLuca-Davis filed a bill in equity.(a) to reform or (b) to rescind its bid. The Court of Appeals refused reformation, but held that DeLuca-Davis was entitled to cancellation of its bid and return of its deposit.

The unanimous opinion of the Court, written by Judge Hammond, analyzed a great many decisions from Maryland and other states and the views expressed in Williston, Corbin, Black, and the Restatement Contracts, and stated:

"Although reformation requires that the mistake be mutual, recission may be granted whether the mistake be that of one or both of the parties." 210 Md. at 526, 124 A.2d at 561.

Most importantly for the purpose of the instant case the Court said: "The general rule as to the conditions precedent to rescission for unilateral mistakes may be summarized thus: (1) the mistake must be of such grave consequences that to enforce the contract as made or offered would be unconscionable; (2) the mistake must relate to a material feature of the contract; (3) the mistake must not have come about because of the violation of a positive legal duty or from culpable negligence; (4) the other party must be put in *statu quo* to the extent that he suffers no serious prejudice except the loss of his bargain." 210 Md. at 527,124 A.2d at 562.

After quoting at considerable length from *DeLuca-Davis* and other Maryland cases, the court applied the tests listed above to the situation before it and came to the conclusion that rescission of the contract was justified; It said:

The College argues that *Kappelman, DeLuca-Davis* and *McGraw* are distinguishable from the instant case because in each of those cases the mistake was palpable, and because in the instant case Miller Inc. was negligent in failing to clarify its confusion or uncertainty with respect to the specifications by asking the architects for an authoritative interpretation.

Taking the latter point first, the Court finds that Miller Inc. was negligent in failing to take the matter up with the architects, and in relying on Miller's construction of the confusing, though not ambiguous specifications; but the Court finds that Miller Inc. was not guilty of gross or culpable negligence. Relief should not be denied because there was "more or less negligence," *Kappelman, supra,* nor because Miller was "to a degree negligent," *McGraw, supra,* since the mistake was not the result of "culpable negligence," *DeLuca-Davis, supra.*

On the question of whether the mistake was palpable, the Court has found as a fact that the mistake was obvious, i.e., legally palpable, to the contractors present, when the bids were opened, and was or should have been obvious to the architects, though not to the other representatives of the college. The Court doubts whether palpability is an essential element of relief in such a case as this; if it is, the college is charged with the knowledge of the architects.

More importantly the college knew of the mistake when Miller Inc. sought to withdraw its bid, before the college had accepted the bid in writing, as it was required to do by the terms of the bid bond before it could hold Miller Inc. or its surety liable thereunder. It is not necessary to decide whether the withdrawal of the bid was such a breach as excused the failure of the college to give written notice of acceptance in view of the continued insistence of the college that Miller Inc. sign a contract on Friday, August 11, a week before it was required to sign by the Instructions to Bidders, the Form of Bid and other relevant documents. The college knew of the mistake in time to be able to sign contracts with the next higher bidders, without loss of the favorable wage scale, for the amounts of their respective bids. The college lost nothing except the bargain which it would have obtained as a result of Miller's bona fide mistake.

The fact that the loss to Miller Inc. would probably not have been so great as to put it out of business is an element to be considered, but it is not controlling. The mistake was substantial and material. . . .

The evidence meets the four tests specified by the Maryland Court of Appeals in *DeLuca-Davis* as "the condition precedent to rescission for unilateral mistakes," quoted above from 210 Md. at 527, 124 A, 2d at 562: The mistake was of such grave consequences that to enforce the contract would be unconscionable; the mistake was material; the mistake did not result from violation of a positive legal duty or from culpable negligence; and there was no change in the *status quo* to the extent that the college suffered any prejudice except the loss of its bargain.

Judgment was entered in favor of the defendants, the surety company and the contractor. (*President and Council of Mount Saint Mary's College vs. Aetna Casualty and Surety Co.*, 233 F. Supp. 787.)

9. Consequences of Missed Deadlines

Nearly every construction contract names a work completion date with the provision that a failure to meet that date will result in a penalty to the contractor. The trio of cases in this section is concerned with the responsibilities for delay and the damages properly awarded under the terms of the three contracts.

When Construction Is Delayed, Who Pays?

Almost all construction contracts contain work deadline clauses and provide penalties for failure to meet the stated deadlines. The provisions, however, usually include escape clauses listing conditions under which the deadline can be postponed. The interpretation of such clauses frequently is decided in court. An excellent example of how these matters are resolved was presented in the United States Court of Claims. A construction firm felt that it had been unfairly penalized for delay in the completion of its contract and brought an action for remission of damages. The court referred the matter to a trial commissioner and then adopted his report as its decision.

The job in question called for the construction of 183 dwelling units in Washington, D.C., for the National Capital Housing Authority, the defendant in the action. The work was not completed within the stipulated time limit and damages were assessed against the contractor firm. It contested the damages for a number of differing reasons.

The general situation was that work was divided into two stages. Each stage was to be "progressively completed, suitable and ready for occupancy" within a prescribed number of consecutive calendar days. Plaintiff was liable for liquidated damages for each calendar day of delay until the work was completed or accepted. Plaintiff was not to be charged with liquidated damages, however, because of any delays due to unforeseeable causes beyond its control and without its fault or negligence, including "unusually severe weather."

The first matter taken up by the commissioner was the delay caused by a paving contractor. In regard to this claim, the commissioner's report said:

> The contracting officer granted plaintiff 130 days extension of contract time for completion of group 1 buildings due to delay caused by a paving contractor, but granted no extension of contract time for completion of group 2 buildings. The testimony of both plaintiff's and defendant's witnesses establishes, however, that the delay of the work on group 1 buildings interfered with the planned sequence of the operations with a resultant delaying effect on the work on group 2 buildings. Plaintiff claims that it should have been granted 70 days extension of contract time for completion of group 2 buildings, but, as set forth in finding 5(f), 15 days extension of time is found to be warranted.

Plaintiff then claimed that it was unduly penalized because of delays caused by bad weather. The contractor firm requested the contracting officer for an extension of contract time because of "extreme weather conditions." The contracting officer, after comparing the weather conditions during the months of the contract work with the weather conditions during similar months of a preceding eight-year period, granted a 22-day extension for each group of buildings. The contractor firm felt that the testimony and exhibits, particularly the government's daily records, indicated the project was delayed by adverse weather conditions for a substantially longer period. The commissioner's report, however, indicated that the evidence adduced by plaintiff regarding weather delays was vague and fragmentary and did not establish that it was entitled to an extension of contract time for a period longer than that allowed by the contracting officer.

Vandalism was another issue raised by the plaintiff. The commissioner's report quoted the pertinent provisions of the contract, and continued:

> In denying plaintiff's request for an extension of contract time because of delays due to vandalism, the contracting officer emphasized that it was not defendant's responsibility under the contract to protect the project from theft or damage while it was in the hands of the contractor, and that there was no requirement in the contract for an extension because of delay due to vandalism. In reaching this conclusion, it is clear that the contracting officer was in error because of the contract provisions which exonerated plaintiff from liquidated damages because

of any delays in the completion of the work "due to unforeseeable causes beyond its control and without its fault or negligence," do not depend upon fault or responsibility of defendant. Moreover, when one couples this language with the contract language which provided that the contractor should not be relieved from payment of liquidated damages because of delay caused by failure of the contractor to adopt reasonable and continuous protective methods, the conclusion is inescapable that if the delay was because of the above unforeseeable causes and was not caused by failure of the contractor to adopt the protective methods, then the contractor was entitled to be relieved from payment of liquidated damages. The contract is, of course, to be interpreted so as to harmonize all provisions whenever it is possible to do so. In preparing its bid, plaintiff anticipated that there would be some vandalism, and during the construction period, plaintiff had two or three watchmen on duty at all times. In addition, plaintiff alerted the Police Department "to keep watch as much as they could." Nevertheless, the evidence is clear and uncontradicted that there was extensive vandalism on the project, including approximately $4,000 to $5,000 in damage to window glass, and that plaintiff was delayed by the vandalism. As set forth in finding 7(d), plaintiff adopted reasonable and continuous protective methods, but was delayed by vandalism which was unforeseeable, beyond its control and without fault or negligence, to the extent of five days on both group 1 and group 2 buildings.

The dispute concerning errors in drawings was determined as follows:

> The contracting officer denied plaintiff's request for an extension of the contract time because of delay which plaintiff asserted was due to errors in the drawings. As set forth in finding 8(b), the weight of the evidence establishes that there were no serious errors in the drawings, and that the corrections in the drawings were handled expeditiously and that the errors in the drawings did not cause plaintiff any significant delay in the over-all progress of the work.

A road passed close to both groups of buildings and required the construction of a retaining wall and an alteration in grade, causing a considerable delay, according to the contractor. The contracting officer agreed that an extension should be made and, after a detailed recital of the facts, said, "Based on the foregoing facts it is my determination as Contracting Officer for the Authority that the Contractor experienced 12 calendar days delay on Buildings A-2 and B-2 and 106 (94 plus 12) calendar days delay on site and landscaping of Group 2 because of unforeseeable causes beyond his control and without his fault or negligence, which delay was not concurrent with other delays for which the Contractor has been given an extension of time. . . . "

However, a claim by the plaintiff for additional time on Group 2 buildings for delays attributable to the same cause was denied.

The plaintiff contended that it was unreasonably delayed because government inspectors failed to inspect the work after the buildings had been substantially completed. After quoting several pertinent provisions of the contract, the commissioner's report quoted from the contracting officer's statement as follows:

> The Authority refused on this project, just as it has on all other projects, to accept any buildings for use and occupancy until they were brought to a satisfactory stage of completion and quality of workmanship to comply with the contract. Numerous punch lists were made on these buildings, and checked and rechecked by the project personnel representing the Architect and the Authority until satisfactory completion of contract work and punch list items. The Authority was constantly pushing the Contractor and his personnel to satisfactorily complete the buildings so that they could be taken over. A review of the memoranda of acceptance for occupancy will reveal that buildings were finally taken over subject to completion of miscellaneous work on exterior of buildings and site work and landscaping, and in some cases completion of interior work. If the Contractor's personnel had been diligent in the satisfactory completion of contract required work and punch list items, the amount of punch listing would have been greatly reduced and the buildings could and would have been taken over sooner than they were. However, they failed to properly check and supervise the mechanics' work, and therefore, any delay is the Contractor's responsibility.
>
> Based on the foregoing facts it is my determination as Contracting Officer for the Authority that the Contractor experienced no delay on the part of the Authority in accepting buildings for use and occupancy that was not caused by his personnel and . . . no time extension is granted.
>
> As to the tenth cause of delay asserted by the Contractor in his letter of February 9, 1962 . . . relating to alleged delay by the Authority in accepting buildings for use and occupancy, and for which an extension of 30 days was requested, I find the facts, as they apply to Group 2, to be as follows:
>
> For the reasons set forth in detail in Item 11 above, the Contractor is entitled to no extension for alleged delays in acceptance of buildings for use and occupancy. . . .

After these quotations from the contracting officer's statement, the commissioner's report summed up the acceptance situation:

At the trial the evidence established that the buildings were "substantially" completed prior to acceptance by defendant. Plaintiff, apparently relying upon the above-quoted Section 34 of the General Conditions, claims that it should have been allowed an extension of the contract time "as a result of the failure of the defendant to inspect and accept the dwelling units when substantially completed." A fair reading of Section 24 of the General Conditions and of the above-quoted paragraph 2 of Section 3 of the Special Conditions, however, compels the conclusion that when the work was substantially completed plaintiff was obliged to notify defendant of that fact, but that defendant was not then obliged to accept the buildings. The language of the Special Conditions is obviously permissive and embraces the exercise of discretion by the defendant as to whether a building was reasonably safe, fit and convenient for the use and accommodation for which it was intended. Moreover the buildings were, as previously indicated, to be "completed, suitable and ready for occupancy." As set forth in finding 10(i), the record does not support plaintiff's contention that defendant's inspectors were tardy in making inspections or that the inspections were unnecessary or improper, or that defendant failed to accept the buildings promptly after they were completed and ready for occupancy.

Finally, plaintiff sought additional recompense for the execution of change orders. The court noted, however, that all of the change orders contained a statement that an equitable extension of contract time would be considered separately. At the trial, defendant's official who prepared the findings of fact for the contracting officer, now deceased, testified that he arrived at an extension of the contract time of 34 calendar days for Group 1 change orders and of 49 calendar days for Group 2 change orders. He calculated these extensions by dividing the contract dollar amount by the contract days. He further testified that after he reached the conclusion he conferred with representatives of the Public Housing Authority (which under the contract approved extensions), and they disagreed with his formula.

The court's opinion continued:

As set forth in finding 11(c), although at the trial plaintiff attacked the above-mentioned formula and on each of several change orders adduced evidence to the effect that a certain number of days were required for a subcontractor to complete the work under the change order, it failed to show how the change order affected either group of buildings or the project as a whole so as to justify additional extension of the contract time. Plaintiff, accordingly, failed to establish that is was entitled to additional time extensions because of change orders.

The commissioner's report concluded as follows:

In its petition, plaintiff alleges that defendant breached the contract by failing and refusing to accept the work when it was substantially completed, and that as a result, plaintiff incurred additional costs; but in view of the conclusion heretofore reached that the record does not support plaintiff's contention that defendant failed to accept the buildings promptly after they were completed and ready for occupancy, this claim for additional costs is not sustained. . . .

In applying the contract formula for liquidated damages and in accordance with the findings of fact on excusable delay, plaintiff is entitled to recover the sum of $4,450. (*Wertheimer Construction Corp. vs. United States,* 406 F. 1071.)

What Is a Reasonable Deadline?

A contract provision specifying the date when contemplated work should be completed is an eminently desirable feature of a construction contract. Unfortunately, circumstances often preclude such a provision's inclusion in a contract. If all goes well, the omission does no harm. But if things do not, and the parties fail to get on with each other, the question of what is a reasonable time for the performance of the work will usually be decided in court.

One such case developed after a Dallas firm of consulting engineers—Leo L. Landauer & Associates Inc.—contracted with Houston County to prepare plans and specifications for the county courthouse's air conditioning system. The engineers were to seek bids on the work from competent contractors and to supervise the work of installation. As it turned out, no funds were made available for the job, and when, after considerable delay, Landauer presented the county with a bill for the preparation of plans and specifications, the county refused to pay it. The engineers recovered a judgment in trial court, but the county appealed to the Court of Civil Appeals of Texas at Tyler.

The appeals court began its opinion by quoting the contract that was embodied in a proposal made by the engineering firm and accepted by the county. It read as follows:

In accordance with our discussion and agreement of September 28, 1959, regarding the proposed air conditioning system for the Houston County Courthouse at Crockett, Texas, we will survey the courthouse, prepare the necessary studies, calculations, plans and specifications for competitive bidding; advertise and submit the bidding documents to recognized contractors; analyze the bids and assist in the preparation of the necessary contracts; check shop drawings and supervise the work of installation for a fee of ten (10) percent of the lowest acceptable bid.

Seventy-five (75) percent of the total fee is attributable to the design stage and is due and payable upon delivery of the final plans and specifications. The remaining is chargeable to supervision and is payable upon completion of the work of installation.

In the event that construction funds have not been appropriated at the time that plans are complete and delivered to you, and you elect not to take bids at the time; or in the event of abandonment or indefinite postponement for any other reason, then the payment of our fee shall be based on construction cost as determined by a qualified estimator.

The appeals court said the engineering firm had brought suit because it had fully and completely performed its part of the contract.

In giving the county's side of the case, the court said, "The appellant answered and denied liability under the terms of the agreement because of appellee's failure to submit the plans and specifications and other bidding documents to recognized contractors as provided in the written agreement and that such breach relieved the appellant . . . of any liability by virtue of the written contract."

After summarizing the provisions of the contract, the court continued:

On Oct. 8, 1959, the County Judge mailed the executed contract to appellee and in the letter of transmittal called appellee's attention to the fact that funds would not be available for the job including appellee's fee (which is the subject matter of this lawsuit) before January 1, 1960, and stated "so you may plan your work accordingly." Appellee met with the Commissioners' Court February 2, 1960, and reviewed the plans and specifications and answered the Court's questions. On March 7, 1960, appellee, in writing, notified appellant that plans and specifications were ready for bidders and that advertising should be done. On March 9, 1960, the County Judge wrote appellee stating that the Court "yesterday refused to advertise the air conditioning job for bids" and, among other things, stated "I certainly think you are entitled to your adjusted 10% engineering fee. . . . "

The appeals court went on to say that on March 11, 1960, the engineers wrote the county urging the completion of the project, but on March 14, 1960, the county judge wrote a letter to the engineers telling them to prepare, at no cost to the county, plans and specifications for a natural gas–powered system instead of an electric system as originally planned. The letter stated that the Commissioners' Court indicated it would accept any bid below $40,000.The appeals court continued:

> On March 24, 1960, appellee by telegram advised appellant "Electric–gas air conditioning plans complete. Advise us when and how advertised." On March 28, 1960, the County Clerk wrote appellee: "The Commissioners asked me to inform you that our County Judge is seriously sick in a hospital, and the Court feels that they must wait until he either gets back to the office, or another judge is appointed. As soon as this problem is solved, we will inform you what to do."
>
> The court never did authorize appellee to proceed after the dual [electric-gas] plans were completed. An officer of Leo L. Landauer & Associates Inc., appellee, testified that appellee never did advertise for bids because "The court told us they would advertise." The court again agreed to advertise for bids on June 11, 1962, but this was never done. Appellee's witness and one of its officers, McCormick, testified that appellant had never indicated any dissatisfaction on account of any delay by appellee prior to the trial of this case. It appears that appellee was ready and willing at all times to perform the obligations imposed upon it under the terms of the contract. A copy of the plans and specifications was physically delivered on or about June 14, 1962, and the jury found this to be the date of delivery. The jury also found that the appellee prepared and completed the plans and specifications in accordance with the written agreement. The jury's finding on this issue was not attacked.

The appeals court pointed out that in the contract no time was fixed for performance, adding:

> It is the settled law of this state that where the contract does not fix a time for performance, the law allows reasonable time for its performance. . . . A reasonable time has been defined to be such time as is necessary, conveniently, to do what the contract requires to be done, and as soon as the circumstances permit. Reasonable time depends upon the circumstances in each case, including the nature and character of the thing to be done and and the difficulties surrounding and attending its accomplishment and is generally for the jury. . . . The difficulty of the performance may be a circumstance to postpone the time. . . .

After a discussion of its right to regard the jury's findings in certain circumstances, the appeals court continued:

> If there is any evidence of probative force, from which reasonable minds might come to the conclusion that the jury came to, then there is "some evidence" to support the finding of the jury, and it becomes the duty of the court to enter judgment in keeping with such finding. The fact that there may be evidence to the contrary is immaterial. . .
>
> After reviewing all the testimony in its most favorable light in support of the verdict as we are required to do . . . it appears to us that there was at least some evidence of probative force to raise an issue of fact upon the question of whether or not such delay by appellee in delivering such plans and specifications was unreasonable delay.

Having conceded that the county might have the semblance of a defense because of the engineering firm's delay of two years in submitting the plans and specifications, the court promptly closed the door once more and found that because of the county's own delays the engineering firm's action was reasonable. On this point it said:

> Appellee admits it did not advertise or submit the bidding documents to recognized contractors as provided for by the terms of the written contract. Appellant contends that this is an admission of a breach of the written contract on the part of the appellee and consequently as a matter of law it cannot recover its fees for services rendered as provided for in said contract. This contention is overruled. In support of its position, it relies upon the rule of law "that a party to a contract who is himself in default cannot maintain a suit for its breach." We recognize this well–established rule of law. However, under the evidence as reflected in this record, we are of the opinion that this case falls within the equally well–established rule of law that "where the obligation of a party depends upon a certain condition being performed, and the fulfillment of that condition is prevented by the act of the other party, the condition is considered as fulfilled."
>
> Witness McCormick testified that at his meeting with the Commissioners' Court on February 22, 1960, the Commissioners' Court told him that they would take care of and do the advertising of these contracts for competitive bidding and that they were going to forego that part of the contract and that he and the Commissioners' Court so agreed. Mr. McCormick further testified that at this meeting, it was pointed out to him that Houston County would take care of the official advertising in accordance

with their normal practice. In view of this testimony, we cannot say that appellee's admission that it did not advertise or submit the bidding documents to recognized contractors establishes, as a matter of law, a breach of the contract.

In accord with its conclusion that by its actions the county justified not only the engineering firm's failure to advertise for bids and submit such bids to competent contractors but also the long delay in delivering the completed plans and specifications, the court affirmed the judgment of the trial court allowing the firm to collect the stipulated fee from the county. Under the circumstances, as disclosed by the record, the court considered the delay to be "reasonable" as that term was defined in its opinion. (*Houston County vs. Leo L. Landauer & Associates Inc.* 424 S.W. 2d 458)

Who Is Responsible for Job Delay?

On the cover of the February, 1967, issue of *Actual Specifying Engineer* there appeared the statement, "The future of the construction industry can be summed up in a word: Awesome." The judge who is called upon to interpret the complicated contracts that are part of the modern construction job with their multiplicity of contractors and subcontractors, in all probabilty is convinced that the hour of awesomeness is already here.

An excellent example of this condition is provided by a memorandum decision of the New York Court of Claims dealing with the perennial question of who is responsible for delays in the progress of a construction job that culminate in the postponement of the completion date, with additional cost to all concerned. Justice Alexander Del Giorno, who wrote the decision, for once did not blame any of the various contractors and subcontractors. Instead, he attributed the delays to the failure of the state to coordinate the work of the different contractors. His memorandum decision follows:

This is a claim against the State by the claimant which was one of seven prime contractors who had bid on several specifications for the erection of the Reception Building, known as Building No. 102, at the Bronx State Hospital, Bronx County, New York. The claimant was to furnish labor and material for the electrical work. Depot Construction Corporation was the general construction contractor. Other contracts were for heating, sanitary work, refrigeration, elevator work and food service equipment.

These contracts were entered into on May 12, 1959, and were to terminate on December 1, 1961. In all the contracts it was provided that the work of each contractor was to harmonize and be installed in conjunction with the work of the other contractors, especially the general construction contractor. The State Architect had supervision of and direction over the contract work as well as authority to bring about proper execution of the contract requirements, particularly work or job coordination, in order to insure unimpeded progress of the contract.

Claimant alleges that it performed all the terms of the contract; that its part of the contract was duly accepted by the State on or about December 20, 1963; that the State breached the said contract in that it unreasonably failed to coordinate the work of other contractors with the work of the claimant, permitted delays in the construction of walls, ceilings, painting, elevators, plastering, hanging ceilings, ceramic tiling, all of which were preliminary to claimant's work and, as a result, claiment's work was hampered and delayed.

The claimant further charges that the contract required the State to provide two hoists for material and one elevator for the workmen. The State, upon removal of hoists, provided only one elevator for hoisting men and materials, all of which caused claimant to expend additional sums of money for work, labor, material and equipment.

The proof indicates that the general contractor was quite indifferent to its obligation to properly man, supervise and coordinate its work and the facilities necessary for the smooth performance of the entire job which held up the work of the claimant and that the State was lackadaisical and indifferent to the many and continuing complaints of the claimant. As an example, concrete slabs for the structure were to be laid commencing February, 1960. They were not laid until the beginning of March, 1960, holding up the related work of all other contractors, including the claimant. Nevertheless, the claimant by maximum effort and even overtime work, finished in time the work involved in that phase of the contract.

The general contractor used one superintendent for three separate jobs on the hospital site and intermingled its operations and men to the detriment of the job on which the claimant worked. The State permitted it to use one engineer for two hoists and also permitted removal of the two hoists when only one interior elevator was available and not at least two, as required by the contract. The brick work was delayed, as well as the lathing, the plastering and the painting, the setting up of door bucks, ceiling work, and the affixing of cabinets.

Hoists were not made available to claimant to the extent required by the contract. Claimant had to store on the first floor over 4,000 fixtures because hoists or elevators were not available, necessitating the use of a crane to lift the fixtures to the various floors, involving thus a double operation of equipment and manpower by the claimant.

The testimony indicated that there were various strikes during 1960 and 1961, which would affect the general operation of the general contract, but from all the evidence it would seem that had the general contractor been made to perform in timely fashion by the State, said general contractor would have been ahead of the several strikes in most instances, as claimed by the claimant. State's finding #134 concedes claimant had performed 90% of its work by June 18, 1962. Although it is difficult to assess how much could have been thus gained or lost by virtue of the strikes, the Court feels that no more than one-quarter of the time lost could or should be assessed in favor of the State in regard to this claim. Likewise, since the claim for time lost waiting for hoists and re-handling of fixtures is an estimate, the Court allows only one-half of the time claimed as a reasonable estimate of that time.

In general, by letters and intermittent oral complaints, the claimant drew the attention of the State to the delays caused by the general contractor, and the answer was almost always that the State could do nothing about these complaints, and that these were merely matters requiring job-site coordination. Behind this "job-site coordination," which is a built-in *a priori* protection, in such contracts the State seemed to take a cavalier, albeit benevolent, attitude towards the many requests of this claimant for use of its effective authority over this contract.

In all, only seven job coordination meetings were held on this job which lasted from May, 1959, to December 20, 1963 and, actually, to about May, 1964, because of many of the small items still to be completed after December 20, 1963.

The claimant's findings found by the Court itemize these delays and complaints, which are to be deemed a part of this decision.

These specify the delays which impeded and held up the claimant's performance of the contract. The State had a duty to coordinate the entire job and failed to perform this duty.

The Court would like to observe that Section 135 of the State Finance Law, and any other statutory requirement, should be studied with the express purpose of either eliminating or amending the law to permit the State to let such contracts as this to one bidder instead of five, six or more bidders, with none having authority over the others but all having the same privilege of screaming for help from the State Engineer on the job, whose own efficiency is diluted because he too often has to "mother" the disputing contractors, rather than perform his primary duty of progressing the job. Experience would indicate that under the prevailing system the State squanders huge sums of money in trying to keep the jig-saw puzzle together, whereas, under the one bid system, the responsibility of efficiency and coordination would not be upon the one contractor but it would be to said contractor's financial advantage to move with coordination, efficiency and due speed to complete the contract, for the basic reason that the contractor could not place upon the shoulders of others, but only upon himself, any blame for a slowdown or uncoordinated work. Were this the case, then the State Engineer would essentially only be called upon to watch the proficiency of the work and not to arbitrate either arguments or uncooperation among many contractors.

An example of the illogical and desultory action taken by the State Architect who depended for his information on the State Engineer on the job is the following letter (Claimant's Ex. 21):

> August 23, 1962
> 32404

Bronx State Hospital
Reception Building #120
Specification 15901 E
Forest Electric Corporation
630 Ninth Avenue
New York 36, N. Y.

Gentlemen:

For the purpose of record and the information of all concerned, this office repeats herewith your August 22, 1962, letter.

We call your attention to the following job conditions:

No elevators are in operation. The lighting fixtures will have to be carried up to the various floors.

The ceilings are not painted throughout the floors. Rooms are skipped. We cannot hang fixtures in one room then move to another floor without completing a given area.

You will agree with us that in order to perform this work in an efficient manner, we have to do the entire floor.

We intend to proceed with the hanging of the fluorescent fixtures in an efficient manner and not be hampered by the inefficiency of the General Contractor.

We also request that steps be taken to provide ample elevator service as called for in the specifications.

The foregoing matter was discussed today with our Local Representative and it is understood that the problem at hand is not a serious one. We would suggest, based on past experience, that all concerned should meet at the jobsite and settle the matter on a mutually agreeable basis in accordance with the terms of the applicable contracts.

Very truly yours,
/s/ C. W. Larsen
State Architect

The Court finds that the delays complained of, except for the portion indicated above to have been affected by the strikes, were not ordinary delays for which the State is liable, the damages for which will be reflected in the Court's decision issued simultaneously with this Memorandum-Decision. (*Forest Electric Corp. vs. State,* 275 N.Y.S. 2d 917.)

10. Arbitration

Arbitration is assuming a prominent place in the construction industry. Today, most construction contracts include provisions for arbitration, and many disputes are settled out of court as a result of implementing such arbitration provisions.

In this final section are four cases of arbitration that *did* go to court.

What Is Arbitration?

In the definition of arbitration taken from the opinion of a Louisiana court that appeared in a discussion of arbitration in the July, 1962, issue of *Actual Specifying Engineer,* the fact that arbitration functions at its best outside the jurisdiction of the courts was emphasized repeatedly. That definition referred to arbitration as a "domestic tribunal" conducted by "unofficial persons" to make awards "in lieu of proceeding for judgment in established tribunals of justice" and called it a "quasi-judicial" means of achieving justice.

This important distinction between arbitration proceedings and litigation in the established courts was reemphasized by a Connecticut court (Superior Court, Hartford County) in a lengthy and carefully reasoned opinion on the question of whether a delay in beginning arbitration proceedings under the pertinent provisions of a construction contract was subject to the six–year statute of limitations that bars the initiation of legal actions in the established courts after the expiration of that period since the right of action accrued. Judge Joseph E. Klau, before whom the matter was argued, ruled that there is a clear distinction between arbitration proceedings and legal actions that permits a greater delay in bringing arbitration proceedings than

is allowed by the statute of limitations in the case of formal legal actions
in the established courts.

It was no trivial dispute that brought the matter into court. The architects of the new home offices of a large life insurance company and the insurance company, found themselves at loggerheads when the heating and air
conditioning systems in the extensive structure failed to function satisfactorily
because of corrosion of the pipes. The entire system had to be rebuilt.

The situation was described by the court:

There is no great dispute with respect to the facts as far as the
present action is concerned. The plaintiff is a widely known partnership
firm engaged in the practice of architecture and on May 5, 1953,
entered into a written contract in this state with the defendant insurance
company whereby the plaintiff agreed to render architectural and supervisory services in connection with the design and construction of the
new home office building of the defendant in the town of Bloomfield.
Under the agreement, the plaintiff was given responsibility for each phase
of the planning and construction, was to prepare all drawings and
specifications for the building, and was given the duty of complete supervision of the course of the "Work." One of the plaintiff's specific
undertakings was to furnish the "services of structural, heating, ventilating, air conditioning, plumbing and electrical engineers." The subject
matter of the controversy here relates to alleged breach of contract with
respect to the heating and air conditioning system.

In February, 1955, the plans for the heating, ventilating and air
conditioning [HVAC] system were "frozen" for the purpose of securing
bids thereof. The purpose of freezing the HVAC plans was so that bids
might be taken from potential subcontractors for the HVAC system on a
uniform basis without the confusion which would result from the necessity of making adjustments in the various bids to take account of design
changes constantly being made.

On March 9, 1955, the plaintiff sent a copy of the HVAC design
and specifications, in their then form, to the defendant. No approval of
these plans from a technical engineering point of view was asked and
none was given, and the plaintiff conceded that the defendant was not
qualified so to approve. The specifications, however, were accepted by
the defendant prior to obtaining bids. Bids were secured from subcontractors, and in May, 1955, the general contractor engaged by the
defendant entered into a subcontract for the HVAC system, and thereafter the general contractor undertook and completed the installation
of the HVAC system, as designed by the plaintiff, in accordance with
the design and specifications furnished by the plaintiff as aforesaid.

While no changes were made after March 9, 1955, in the basic specifications for the pipe which subsequently corroded and thereby precipitated the underlying dispute between the parties, many changes
were made in plans of the HVAC system and in other aspects of the
building subsequent to this date. In fact, changes thereafter made in
the HVAC system resulted in an additional cost of $500,000.

In the early months of 1957, the defendant occupied said building as a home office and has continuously occupied it since that time.
The defendant first became aware of the defective air conditioning
system shortly before June 10, 1960. On that date, the defendant
notified the plaintiff that there was evidence of serious pipe corrosion in
the heating and air conditioning system which used, as a cooling
agent, well water from ten wells constructed on the premises of the
defendant. On that date and thereafter, meetings were held to determine the cause of this corrosion and the remedial steps which should
be taken. A representative of the plaintiff attended several of these
meetings, and the plaintiff received copies of the minutes of all of
them.

The actual design of the HVAC system had been undertaken
by well-known mechanical engineers, Syska and Hennessy, selected
by the plaintiff and approved by the defendant. This was in accordance with the terms of the agreement entered into between the
plaintiff and the defendant. The pipes specified and actually installed for the heating and air conditioning system were copper
coils, and the corrosive action thereof was in all probability due to
the chemical content of the water used in connection with the
heating and air conditioning system and obtained from the wells
driven for this purpose. The plaintiff and the firm of Syska & Hennessy,
the plaintiff's subcontractor who actually designed the HVAC system,
both obviously familiar with the building, were retained by the defendant
to devise a means of altering the HVAC system in light of the corrosion
problem. Changes in the system were subsequently made.

On July 31, 1962, defendant demanded, pursuant to the
arbitration clause contained in the contract of May 5, 1953, that
plaintiff arbitrate the question of whether the corrosion in the
HVAC system was due to a breach by plaintiff of its duties under
this contract.

The court then quoted the arbitration provision of the contract
which called for submission of questions in dispute "in accordance with
the procedures then obtaining of the American Arbitration Association,"
and continued by stating the positions of the opposing parties in regard
thereto. These positions were:

The plaintiff architectural firm refused to arbitrate, basing its refusal on a contention that the arbitration proceedings were barred by the
Statute of Limitations. It contended that the cause of action, if any, arose
in March, 1955, when the plaintiff submitted the HVAC specifications to
the defendant and the defendant approved them. As more than six years
had elapsed between March, 1955, and July, 1962, the architectural firm
argued that all proceedings were barred.

The defendant insurance company's defense was more elaborate. It
contended that first, the Statute of Limitations did not apply because an
arbitration proceeding is not an "action" as defined in the statute; second,
that even if it were, the question of its applicability should be left to the
arbitrators and not to the court; third, that even if the defendant's position on the two points above should not be upheld, the breach of the
contract did not occur on a specific date in 1955 but that the plaintiff's
responsibilities under the contract included supervisory services which
lasted until the completion of the building in 1957, thus eliminating the
six-year limitation, as the demand for arbitration was made within six
years of the date the building was completed.

The court disposed first of the questions of whether or not
arbitration proceedings are actions within the definition of the Statute
of Limitations and ruled that they are not. On this point it cited several decisions from Connecticut and other jurisdictions, saying:

> Arbitration is not a common law action, and the institution
> of arbitration proceedings is not the bringing of an action under any
> of our statutes of limitation. "Arbitration is an arrangement for taking
> and abiding by the judgment of selected persons in some disputed
> matter, instead of carrying it to the established tribunals of justice;
> and is intended to avoid the formalities, the delay, the expense and
> vexation of ordinary litigation. . . ." While it is perfectly true that a
> court proceeding may arise which is related to an arbitration, as when
> a party to an arbitration applies to the court for confirmation of an
> award of the arbitrators, nevertheless, there is no reason to confuse the
> arbitration with legal proceedings which may follow. Any suit to enforce
> an arbitration award is based not upon the original cause of action
> giving rise to the dispute between the parties but upon the award of
> the arbitrators as such.

Having thus drawn the line between a legal action in an established
judicial tribunal and arbitration proceedings, and supporting the defendant
insurance company's position on that point, the court considered the defendant's second contention: that, in any event, the decision should be left to
the arbitrators, and not to the court. Again the court ruled in favor of the

defendant insurance company. It cited a number of authoritative decisions from various jurisdictions, and said:

> Even if it be assumed by way of argument that the commencement of an arbitration proceeding is the bringing of an action within the meaning of the phrase as used in the Statute of Limitations, the application of the statute is to be determined in the arbitration proceedings by the arbitrators and not by the court. The arbitration clause in paragraph 1 of the contract between the parties hereinbefore set forth, is unrestricted in its scope. It leaves all questions in dispute, under the contract, at the choice of either party to arbitration under the procedures then obtaining of the American Arbitration Association. There is no great public policy involved which should lead the court to restrain the defendant from demanding arbitration. . . .

The court also ruled in favor of the defendant insurance company's third contention. It held that the fact that the plaintiff architectural firm was obligated by the contract to continue its supervisory services until the completion of the building extended the time during which an action could be brought until six years after the date of completion, and that the defendant's demand for arbitration came well within the six-year period after the structure was finished and turned over for occupancy to the insurance company.

This carefully prepared opinion involving so important a construction job should be valuable as a clarification of the nature of arbitration proceedings, and the manner in which they differ from ordinary lawsuits in the regular courts. (*Skidmore, Owings & Merrill vs. Connecticut General Life Insurance Co.,* 197 A.2d 816)

How Is Arbitration Carried Out?

Some of the basic principles which guide the courts in cases concerned with arbitration proceedings in construction disputes were restated by the Court of Appeals of Maryland. The award made by a Board of Arbitrators formed to settle a dispute involving the construction of a fire house was challenged by the losing party. When the trial court affirmed the decision of the board, an appeal was carried to the Court of Appeals.

The court's opinion stated the facts of the dispute as follows:

> On April 30, 1962, Button & Goode [contractor] and Chillum-Adelphi [owner] entered into a construction contract whereby Button & Goode agreed to erect two buildings for Chillum-Adelphi.

Plans and specifications had been drafted by the owner's architect, Philip W. Mason. The arbitration proceedings and this suit are concerned with one of the two buildings, the other having been fully completed as required by the contract.

Article 2 of the construction agreement provided that work to be performed under the contract was to commence upon written notice; and the building was to be substantially completed 180 calendar days from the date of such notice. Article 45 of the American Institute of Architects' General Conditions of Contracts, made part of the construction agreement in this case by Article 1 of that agreement, provided that the time in which the contractor agreed to complete the work was of the essence of the contract, and failure to complete the work within the time specified would entitle the owner to deduct—as liquidated damages—out any money which may be due the contractor under the contract, the sum of $50.00 for each calendar day in excess of the 180 days until the building should be substantially completed.

The owner's architect specified that one of the buildings was to be constructed of pre-cast concrete framing. Button & Goode could not commence work until that material was delivered to the building site, and the long and protracted delay of Nitterhouse Concrete Products, Inc. (Nitterhouse) in delivering the concrete frames caused a delay in completing the building beyond the 180 days agreed upon as the time within which construction was to be substantially completed. Chillum–Adelphi retained $21,426.48 of the contract price as damages occasioned by Button & Goode's delay in substantially completing the building.

Article 40 of the General Condition of Contracts provided that the owner and contractor would submit all disputes, claims or questions arising under the contract to arbitration under the procedure then obtaining in the Standard Form of Arbitration Procedure of the American Institute of Architects (AIA). Button & Goode filed a demand for arbitration with the American Arbitration Association (AAA). Chillum–Adelphi objected to the arbitration procedure provided by the AAA; however the parties agreed to submit their dispute to arbitration by the AAA provided that the procedure complied with that of the AIA whereby the parties would be given the opportunity to examine and cross–examine all witnesses and introduce exhibits at any time during the hearing.

It was agreed between Button & Goode and Chillum–Adelphi that the issues to be decided by the board of arbitrators would be: (1) what damages, if any, should be assessed against the contractor in this case, and (2) was the building completed at the time of arbitration.

After delineating the facts which gave rise to the dispute, the court continued by telling what the Board of Arbitrators did about them:

> A hearing was held by the board of arbitrators on August 26, 1964, The arbitrators found that the owner's architect had specified that precast concrete material of Nitterhouse's manufacture be used in the construction of the building, that the contractor had made repeated attempts to have some other company substituted for Nitterhouse to supply the pre-cast concrete frames, but the architect refused to authorize a change because he expected delivery from Nitterhouse sooner than from another company since the order had been pending there for such a long time. Furthermore, a change of suppliers would have necessitated a change in the plans of the building.
>
> Article 18 of the General Conditions provided that the owner's architect should extend the time for the completion of the building if the contractor be delayed in the progress of the work "for any cause beyond the contractor's control." The arbitrators found that Chillum-Adelphi was bound by the decision of its agents, its architect Mr. Mason, to use a product in the construction of the building which proved to be unavailable. The contractor was therefore not responsible for any delay in construction until January 11, 1963, the date Nitterhouse delivered the concrete frames. Under the circumstances, the delay was "beyond the contractor's control" and the architect should have extended the time for completion of the job.

After ruling that the owner was responsible for the delay, the arbitrators proceeded to calculate the amount due the contractor. As the contractor had required 211 days to complete the building after the date the frames were delivered, damages for 31 days at $50 per day, amounting to $1,550, was deducted from the amount withheld by the owner. The sum of $19,876.48 was awarded to the contractor and the costs were equally divided between the parties.

The contractor, Button & Goode, then filed a petition for judgment on the arbitration award and the circuit court entered summary judgment for $20,591. The owners, Chillum-Adelphi, opposed summary judgment on the theory that the arbitrators went beyond the issues submitted to them for determination, that no evidence existed to support the arbitrators' findings, and that a genuine dispute existed because of the fact that the hearing was not conducted in accordance with the Maryland rules.

The court ruled that Maryland's Uniform Arbitration Act did not apply to the case before it, because the act provides that it is not to apply to agreements made prior to June 1, 1965. The Court of Appeals then discussed the general principles underlying arbitration proceedings:

> An arbitration award is the decision of an extra–judicial tribunal "which the parties themselves have created, and by whose judgment they have mutually agreed to abide." [Citation] When suit is brought to enforce the award, a court will not review the findings of law and fact of the arbitrators, but only whether the proceedings were free from fraud, the decision was within the limits of the issues submitted to arbitration, and the arbitration proceedings provided adequate procedural safeguards to assure to all parties a full and fair hearing on the merits of the controversy [Citations] . . . .
>
> Although a court may modify an arbitration award for a mistake in form such as an evident miscalculation of figures (Maryland Rule E4g), an arbitrator's honest decision will not be vacated or modified for a mistake going to the merits of the controversy and resulting in an erroneous arbitration award, unless the mistake is so gross as to evidence misconduct or fraud on his part. [Citation]
>
> In short, where parties have voluntarily and unconditionally agreed to submit issues to arbitration and to be bound by the arbitration award, a court will enter a money judgment on that award and enforce their contract to be so bound unless notwithstanding that the arbitrator's decision may have been erroneous, the facts show that he acted fraudulently or beyond the scope of the issues submitted to him for decision, or that the proceedings lacked prodedural fairness. A court does not act in an appellate capacity in reviewing the arbitration award, but enters judgment on what may be considered a contract of the parties, after it has made an independent determination that the contract should be enforced. . . .
>
> We hold on this appeal that the trial court properly granted Button & Goode's motion for summary judgment on the arbitration award.

Taking up the owner's objections one by one, the Court of Appeals continued:

> There is no merit in Chillum–Adelphi's contention that the arbitrators went beyond the issues submitted to them for determination. Chillum–Adelphi and Button & Goode had agreed that one of the issues to be submitted to arbitration was what damages, if any, should be assessed against the contractor in this case. The architect had refused to extend the time for completing the construction since he did not

feel that the delay was occasioned by circumstances beyond the contractor's control. Since the gravamen of the arbitration proceedings was the fact that because of a delay in completing the contract, Chillum–Adelphi had withheld monies otherwise due the contractor, the arbitrators were clearly authorized to determine whether the architect was correct in his determination that the time for completing the contract should not have been extended. It was essential to review the architect's decision before a determination could be made as to what damages, if any, would be assessed against the contractor.

Chillum–Adelphi's second contention is likewise without merit. The fact that arbitrators may fail to follow strict legal rules of procedure and evidence is not a ground for vacating their award. [Citation] The procedure followed at the arbitration hearing was fair and in full compliance with the AIA procedural rules which the parties agreed would govern the determination of their dispute. The record in the arbitration proceedings remained open for a full six months before the final award was entered. Additional evidence could have been presented to the arbitration board at any time during that six month period, and upon good cause shown the hearing could have been reopened.

Finally, we must discount Chillum–Adelphi's bald assertion that the determination of the arbitration board was unsupported by the evidence. There is no showing of lack of good faith or fraud on the part of the arbitration board, and we will not review the award on the merits.

The underlying facts in this case are undisputed. Since the only inference reasonably to be drawn from these facts was that the arbitration proceedings were fair and that the board did not exceed its powers, no issue of fact existed to be tried in the court below. Button & Goode was entitled to judgment on the arbitration award as a matter of law.

The judgment of the trial court in favor of the contractor was affirmed. (*Chillum-Adelphi Volunteer Fire Department vs. Button & Goode Inc,.* 219 A.2d 801.)

Arbitration or Engineer's Decision, Which Takes Precedence?

A conflict between two strongly worded provisions of a construction contract presented a difficult task to the Court of Appeals of Arizona, Div. 2. It had to decode which of the seemingly irreconcilable views should prevail. The plaintiff in the action, a contractor, contended that a contract provision calling for arbitration of "all questions or controversies" should take precedence over a provision making the engineer "sole judge of the amount due and payable under the contract." The trial court had ruled in favor of the defendant owner, but the contractor appealed the decision.

The appeals court's opinion, including a thorough discussion of the problems involved, contained the following facts:

On Sept. 25, 1967, the owner (Lake Patagonia Recreation Assn.) and the contractor (New Pueblo Constructors Inc.) entered into a contract for the construction of a dam on Sonoita Creek in Santa Cruz County, Ariz.

Paragraph 26 of the contract provided that all questions or controversies between the contractor and the owner, in reference to the contract, should be subject to the decision of some competent person chosen by the owner and the contractor. This person's decision would be final and conclusive for both parties. If the owner and contractor were unable to agree upon who this person should be, a board of three arbitrators would be chosen, one by the owner, one by the contractor, and the third by the chosen two. The decision of any two of the arbitrators would be final and binding upon the parties.

In the fall of 1968, a dispute arose as to whether the dam had been completed by New Pueblo within the time specified by the contract, and whether Lake Patagonia was entitled to liquidated damages. In addition, the contractor contended that the project engineer had miscalculated certain quantities of material supplied by New Pueblo to the project and, therefore, miscalculated the amount of money owed the contractor.

On Jan. 29, 1969, the engineer certified that the work covered by the contract had been completed and accepted. He certified that $81,972.52 was owed the contractor. New Pueblo, however, filed a mechanics lien on Jan. 10 and 17, 1969, claiming that more than $124,800 was due.

On April 18, 1969, New Pueblo, pursuant to Paragraph 28 of the Contract, sent a letter to Lake Patagonia requesting arbitration of the disputed amount due.

In the first paragraph of that letter, notice was given to Lake Patagonia to "designate an arbitrator . . . to participate in the decision of the following dispute: payment of the claim of New Pueblo Construction Inc. of the sum of $124,837.02 based upon proper measurement and computation of quantities and for completion of all items required by the project engineer, the designated agent of the owner, to be done by the contractor."

On April 21, 1969, Lake Patagonia filed a complaint in the Santa Cruz County Superior Court. Two days later counsel for the owner refused to concur with the request for arbitration.

On April 25, Lake Patagonia filed an application in the superior court for an order to stay arbitration. New Pueblo filed a response to the application to stay arbitration and affirmatively moved for an order to compel arbitration. The contractor also moved the court for an order staying the action filed in the superior court.

The court granted the stay of arbitration and refused to enter an order requiring arbitration.

In his appeal, the contractor presented for review the following questions:

Whether the trial (superior) court erred in denying New Pueblo's motion for an order compelling arbitration; erred further in denying New Pueblo's motion to stay the action pending arbitration; and further erred in granting Lake Patagonia's application to stay arbitration.

Lake Patagonia contended that these questions should be answered in the negative because the trial court retained jurisdiction over issues unrelated to the demand for arbitration, and the matter that was raised, and the demand of New Pueblo, was not subject to arbitration.

The appeals court said:

> It is Lake Patagonia's contention that the only matter as to which New Pueblo has sought arbitration by virtue of the letter of April 18, 1969, is the amount of money which New Pueblo claims to be due. It therefore reasons that any determination in the case as to the arbitrability of that issue has nothing to do with its quiet title action, the right of Lake Patagonia to sue for damages for malicious filing of the liens and its right of liquidated damages resulting from New Pueblo's failure to complete the work covered by the contract. It contends that since none of these issues is within the demand for arbitration, their resolution of said issues is completely beyond the authority of any arbitrator. Lake Patagonia cites no authority for this contention, and we have found none.

The appeals court then quoted Arizona's supreme court on the theory of arbitration:

> Broadly speaking arbitration is a contractual proceeding, whereby the parties to any controversy or dispute, in order to obtain an inexpensive and speedy disposition of the matter involved, select judges of their own choice and by consent submit their controversy to such judges for determination in the place of the tribunals provided by the ordinary processes of law. (*Gates vs. Arizona Brewing Co,.* 54 Ariz. 266, 95 P. 2d 49 [1939].)

Using this statement as the basis for its decision, the appeals court decided that arbitration clauses should be construed liberally and that any doubts of whether the matter in question is subject to arbitration should be resolved in favor of arbitration. The court also pointed out that federal courts have adopted the "positive assurance" test that requires arbitration unless it can be said with "positive assurance" that the arbitration clause does not cover the dispute.

In the present case, the appeals court said, the parties had agreed that all questions or controversies between the contractor and owner, under or in reference to the contract, would be subject to the decision of some competent person to be agreed upon by the owner and the contractor. The appeals court continued:

> We think it can be clearly said that all matters raised in Lake Patagonia's complaint are matters which have their basis in the contract. Lake Patagonia's contention that the trial court still has jurisdiction because the matters set forth in its complaint were not covered in the demand letter of April 18, 1969, is entirely without merit. The only matter that can fairly be said not to be within the letter is Lake Patagonia's claim for liquidated damages as the result of New Pueblo's alleged failure to complete the work on time. New Pueblo's motion in the court to require arbitration of the matters set forth in Lake Patagonia's complaint is sufficient demand for arbitration of the matter of the right to liquidated damages.

Lake Patagonia had contended that the matters set forth in its complaint in the trial court were not subject to arbitration because of the fact that the contract excludes decisions of the engineer from arbitration. This contention was based on Paragraphs 14 and 20 of the contract.

Paragraph 14 stated that "The Engineer shall have full authority to interpret the Plans and Specifications and shall determine the amount, quality and acceptance of the work and supplies to be paid for under this contract and every question relative to the fulfillment of the terms and provisions therein. . . ." Paragraph 20 said:

> Partial payments will be made as the work progresses at the end of each calendar month, or as soon thereafter as practicable on estimates made by the Engineer and as approved by the Owner, provided that the Contractor is performing the overall job in a diligent manner. In making partial payments, there shall be retained 10 per-' cent on the amount of each estimate until final completion and acceptance of all work covered by the contract. Upon completion and acceptance of the work, the Engineer shall issue a certificate that the work has been completed and accepted by him under the conditions of this contract, and shall make and approve the final estimate of the work. The entire balance found to be due the Contractor, including

the retained percentages, but excepting such sums as may be lawfully retained by the Owner, shall be paid to the Contractor. Such payment shall be conditioned, however, upon the submission by the Owner that all claims for labor, material and any other outstanding indebtedness in connection with this contract have been paid.

The appeals court's opinion continued:

Lake Patagonia takes the position that these paragraphs make engineers the sole judge of the amount due and payable under the contract, that his judgment is therefore final and conclusive and not subject to the arbitration clause. The type of provisions as provided in Paragraphs 14 and 20 of the contract at issue have been held as making the engineer the sole and final judge of the quality, quantity and acceptability of the contractor's work. [Citations]

If it were not for other provisions in the contract, we would have to agree with Lake Patagonia that the questions involved have been excluded from arbitration. However, the contract must be construed so that every part of it is given effect. [Citation]

The courts must harmonize all parts of the contract, according to the appeals court's view, and conflicting provisions will be reconciled by an interpretation that views the entire instrument. The court cited paragraph II, B.2, Page 3 of the special conditions of the contract, which read:

The Owner's representative and the Contractor's representative shall be present during the classification of material excavation. Upon written request of the Contractor a statement of quantities and classification of excavation in designated localities will be furnished the Contractor within 10 days of the receipt of such request. The statement shall be considered as satisfactory to the Contractor unless specific objections thereto, with reasons therefor, are filed with the Owner, in writing, within 10 days after receipt of said statement by the Contractor or his representative on the work. Failure to file such written objections within said 10 days shall be considered a waiver of all claims based on alleged erroneous estimates of quantities or incorrect classification of materials for the work covered by said statement.

The court ruled out the view that the engineer's decision was final and not subject to arbitration, saying:

This paragraph indicates to us that it was never intended that the decision of the engineer would be final and conclusive on the matters that are at issue. Why give the contractor a right to a claim based on alleged erroneous estimates of quantities or incorrect classification of materials if the decision of the engineer was to be final? Furthermore, the contract provides that it is the duty of the engineer to enforce the specifications in a fair and unbiased manner and that computation of quantities of material that are used shall be made by the engineer in accordance with the methods defined in the plans and specifications. If the engineer fails to perform his duty in a fair and unbiased manner or fails to compute the quantities in the method defined in the plans and specifications, is his decision still to be final and conclusive? We think not. Even the cases cited by Lake Patagonia for the proposition that the engineer's decision is final and conclusive except acts on the part of the engineer that are arbitrary, capricious or unreasonable. New Pueblo did file an objection pursuant to the provisions of Paragraph II, B.2 of the special conditions of the contract and such objection was rejected. The whole tone and tenor of the claim of New Pueblo is that the action of the engineer was arbitrary, capricious and unreasonable.

The court said it believed that the arbitration clause nullified any finality with which the engineer may be cloaked by Paragraphs 14 and 20 of the agreement, except those matters upon which the parties made the decision of the engineer conclusive, as found in Paragraph 9, which stated, "The Contractor shall not take advantage of any errors, discrepancies or omissions which may exist in the Plans and Specifications, but shall immediately call them to the attention of the Engineer whose interpretation or correction thereof shall be conclusive."

The court said that if it were to follow the contention of Lake Patagonia as to the finality of the engineer's decisions, it would have to write the agreement to arbitration completely out of the contract, since there would be nothing left to arbitrate. This the court declined to do.

Lake Patagonia had also contended that the matters which would be decided by the arbitrators would involve both questions of law and fact, and that questions of law are to be decided by the courts and not by arbitrators. This argument, too, was without merit, according to the court, since "an arbitration agreement such as the one in this case gives the arbitrators full power to decide both question of law and fact."

The court devoted a paragraph to Lake Patagonia's contention that a clause in the contract providing for approval of the Farmers Home Administration would make it necessary for that agency to approve any awards made by the arbitrators. Again the court disagreed, pointing out that such clauses did not make federal agencies "the final judge as to whether or not either party had performed under the contract."

In conclusion, the court said:

Lake Patagonia contends that the $14,173.24 claimed by New Pueblo to be due is for extra work which was not approved by the engineer. Because of this Lake Patagonia claims that such extra work did not arise "under or in reference to" the contract. We do not agree. We believe that it does arise under or in reference to the contract. Paragraphs 10 and 11 of the contract relate to extra work and charges and claims for extra costs. The fact that the engineer would not issue any contract change orders and would not approve of the claim does not preclude New Pueblo from asserting its claim. For the requirement of a change order as a condition precedent to recovery will be eliminated in the presence of gross mistake, fraud or error amounting to a failure to exercise an honest judgment. [Citation] This issue really involves the merits of the action which are not the concern of the courts. [Citation]

In its last dying volley, Lake Patagonia claims that in any event none of these issues is arbitrable because the arbitration clause only refers to controversies between the contractor and the owner and this controversy is between the engineer and the contractor. This contention is spurious. The agreement makes the engineer the agent of the owner.

For the foregoing reasons the order of the trial court is set aside, and the case is remanded to the trial court for further proceedings consistent with this opinion.

It seems hardly necessary to add that in the struggle between the arbitration clause and the claim purporting to make the engineer's decisions final and conclusive, the arbitration clause was an easy winner. (*New Pueblo Constructors Inc. vs. Lake Patagonia Recreation Assn. Inc.,* 467 P.2d 88.)

Can a Contract Limit Arbitration?

The Supreme Court of Pennsylvania handed down a decision interpreting the arbitration clauses of a contract. In this case, some eleven months after the contractor had accepted a check for final payment for its work for the construction of an additional wing to a hospital, it sought to reopen a dispute through arbitration. The hospital association thereupon filed a suit in equity to enjoin the contractor from seeking arbitration. The Court of Common Pleas, Westmoreland County, rendered judgment for the hospital association and the contractor appealed to the Supreme Court of Pennsylvania.

That court's opinion, written by Justice Musmanno, follows in full:

On March 16, 1964, the Westmoreland Construction Company, hereinafter called the contractor, entered into a contract with the Westmoreland Hospital Association to erect an additional wing to the Westmoreland Hospital. During the performance of the construction work, the contractor complained that it was encountering delays in its operation because effective liaison between the hospital staff, the architectural firm and the contractor, was lacking. The contractor endeavored to correct the averred fault but was unsuccessful in its efforts. Because of the trouble here referred to, the contractor incurred an alleged loss of $37,203.

On May 19, 1965, the contractor filed a notice of demand for arbitration with the American Arbitration Association of Pittsburgh charging that, because of failure of cooperation on the part of the Westmoreland Hospital Association, it had incurred losses in the amount of $37,203. The Hospital Association then filed a Complaint in Equity in the Westmoreland County courts asking that the contractor be enjoined from seeking arbitration of the indicated dispute. The Court of Common Pleas did so enjoin the contractor, and the contractor has appealed.

The court below in its Opinion, stated that the arbitration clause in the contract was not applicable, because that clause applied to the corpus of the contract and not to any collateral controversy that might arise out of the performance of the contractual enterprise. The court here was in error because the agreement specifically declared:

"7. Arbitration—(2) It is mutually agreed that all disputes arising in connection with this contract shall be subject to arbitration." [Emphasis added]

This language is sufficiently broad to embrace differences between the parties over expenditures and losses caused by delays in the execution of the contract that is to say, "in connection" with the contract. [Citation]

If there had been nothing else in the contract regarding arbitration than the sentence above quoted, the contractor could walk unmolestedly into the arbitration tribunal and demand the summoning of its recalcitrant co-contractee. But there is something else in the contract. Indeed in preparing the contract, the contractor built two barricades before the arbitration tribunal through which it cannot now pass or successfully surmount. These two barricades are the following provisions:

"(b) The work under this contract shall not be stopped or delayed in any way during the arbitration proceedings except by written mutual consent of both parties to the contract, and such mutual consent shall stipulate whether extension of the time for completion of the contract will be authorized by such stoppage or delay.

"(c) Demand for arbitration in connection with any dispute shall be filed in writing by the architect and with all the other parties to the contract. Any demand for arbitration shall be made within thirty days after arisen if practicable, but, in any event, no demand for arbitration shall be made after final payment except in the case of a dispute arising in connection with any guarantee provisions of the Contract Documents."

Obviously, the purpose of clause (b) was to accelerate resolution of disputes during the course of construction so that the construction job could proceed with the least interruption possible. The purpose of clause (c) was to assure the disposition of all claims and the liquidation of all incurred expenses prior to final payment so that, once the final check passed into the hands of the contractor, there would be no unresolved claims hovering over the completed edifice to embarrass and harass its occupants or impede the full enjoyment thereof.

In spite of the clarity of the language which spelled out the need for a written consent to extend the time for completion of the work, in the event of arbitration being required due to disputes, the contractor failed to obtain such a stipulation.

In spite of provision (c) which declared with the specificity of a train or airplane schedule, the final date on which a claim for arbitration could be made, the contractor waited until that final date had passed before it sought arbitration.

When the contractor noted that others were causing delays it could have at once called for arbitration. It failed to do so. With arbitration proceedings being contemplated, the contractor accepted final payment. By its own action, it made impossible arbitration proceedings under the contract.

The Court below properly decreed the injunction prayed for. Decree affirmed; each party to bear its own costs. (*Westmoreland Hospital Association vs. Westmoreland Construction Company,* 223 A.2d 681.)

Index

tion vs. *Davis and Floyd Engineers,* 108–11

"Specification for Design, Fabrication and Erection of Structural Steel for Building," 116

Specifications. *See* Plans and specifications.

"Standard Form of Arbitration Procedure" (AIA), 216, 219

State vs. Board of County Commissioners of Montgomery County, 27–30

State vs. Durham, 136

State vs. Montgomery, 126–28

State vs. T.V. Engineers of Kenosha Inc., 141–43

Statute of limitations, 211–12, 214–15

Stepp vs. Renn, 168

Strict compliance
  in bidding, 182
  with plans and specifications, 2–44

Strikes, 70, 208, 210

Substantial completion, 200–2

Substantial compliance, 41–42
  defined, 23–27

Supervision
  architect's duty in, 103–4
  as art, 83–88
  of construction, 11, 15
  definition of, 91–92
  failure of, 85
  negligent, 116
  responsibility for, by engineer, 83–88, 115–20
  as word of art, 87–88
  as right, 117

Supervisory
  authority of owner, 165–67
  duty of architect, 120
  relationships, statutory, 90–91
  responsibility, extent of, 98–104
  responsibilities of engineer, 165–70

services, payment for, 121–22

rights of owner, 119

Taxpayer's suit, 27–28, 113

Technical specifications, governing, 19

Technical supervision, method of installation, 89

Testimony, expert, 87–88

Town of Poughkeepsie vs. Hopper Plumbing and Heating Corp., 13–18

Trade usage, 38–39

Unilateral mistake, 19–20, 187–88, 195

Unions, labor, 177–78

Unreasonable economic waste, 40–42

Vandalism, 198–99

Verich vs. Florida State Board of Architecture, 123–26

Walker vs. Wittenberg, Delony & Davidson Inc., 170–74

Wallach vs. Salkin, 121–22

Warranty of fitness, implied, 162–65

Weinrott, 149–50

Wertheimer Construction Corp. vs. United States, 197–202

Westinghouse Electric Corp. vs. NLRB, 88–93

Westmoreland Hospital Association vs. Westmoreland Construction Company, 225–27

Willner et al. vs. Woodward, 101

Wilson vs. Village of Forest View, 143–46

Work
  extra, 74, 75
  supplementary, 72
  value of, 76–81

Workmen's compensation insurance, 155

Workmanship, defects in, 63–68